EXPORT DEPENDENCE VERSUS THE NEW PROTECTIONISM

Constraints on Trade Policy in the Industrial World

GLENN RANDALL FONG

Routledge
Taylor & Francis Group
LONDON AND NEW YORK

First published in 1996 by Garland Publishing, Inc.

This edition first published in 2018
by Routledge
2 Park Square, Milton Park, Abingdon, Oxon OX14 4RN

and by Routledge
711 Third Avenue, New York, NY 10017

Routledge is an imprint of the Taylor & Francis Group, an informa business

British Library Cataloguing in Publication Data
A catalogue record for this book is available from the British Library

ISBN: 978-1-138-06323-5 (Set)
ISBN: 978-1-315-14339-2 (Set) (ebk)
ISBN: 978-1-138-30516-8 (Volume 11) (hbk)
ISBN: 978-0-203-72957-1 (Volume 11) (ebk)

Publisher's Note
The publisher has gone to great lengths to ensure the quality of this reprint but points out that some imperfections in the original copies may be apparent.

Disclaimer
The publisher has made every effort to trace copyright holders and would welcome correspondence from those they have been unable to trace.

EXPORT DEPENDENCE VERSUS THE NEW PROTECTIONISM

Constraints on Trade Policy in the Industrial World

GLENN RANDALL FONG

GARLAND PUBLISHING, Inc.
New York & London / 1996

Library of Congress Cataloging-in-Publication Data

Fong, Glenn Randall.
Export dependence versus the new protectionism : constraints on trade policy in the industrial world / Glenn Randall Fong.
p. cm. — (Foreign economic policy of the United States)
Includes bibliographical references and index.
ISBN 0-8153-1599-6 (alk. paper)
1. Protectionism. 2. Protectionism—United States. 3. United States—Commerce. 4. Machine-tool industry—United States—Case studies. I. Title. II. Series.
HF1713.F646 1996
382'.73'0973—dc20 95-53938

Printed on acid-free, 250-year-life paper
Manufactured in the United States of America

to my mother,

Esther,

and the memory of my father,

Sidney

Contents

Figures

Tables

Acknowledgments

The debts I have accrued during the course of this work, while perhaps few in number, are qualitatively pervasive. I thank Professor Richard Rosecrance for providing the intellectual inspiration for this study. His support and encouragement were absolutely instrumental in seeing this work to its proper conclusion. I am grateful to Professor Peter Katzenstein for instilling an analytical rigor, conceptual preciseness, and empirical underpinning in this work. I am also grateful to Professors Patricia Leeds, T. J. Pempel, and Judith Reppy for their comments on all or parts of the manuscript.

The research for this study was supported by grants from the Peace Studies Program, the Center for International Studies, and the Government Department, all of Cornell University. The cooperation of the National Machine Tool Builders' Association is also gratefully acknowledged. The preparation of the final manuscript was supported by the American Graduate School of International Management.

This book is dedicated to my parents, Sidney Fong and Esther Fong Maynard. They encouraged my studies and academic ambitions, no matter what form or shape they took. Any achievement represented by this book is only a humble attempt to follow through on the broader values of excellence and integrity inspired by my parents.

Export Dependence Versus the New Protectionism

Chapter I

Export Dependence as a Constraint on Protectionism

INTRODUCTION

In an international political economy characterized both by constancy and change, this study links together one seemingly incongruous continuity in international trade relations with an increasingly dramatic development in the economies of industrial countries.[1] On the one hand, industrialized economies have become progressively dependent upon one another. Numerous studies have demonstrated that the ratio of trade to national production has steadily increased since the 1960s for all industrial countries.[2]

On the other hand, despite scholarly expectations and real world pressures to the contrary, the liberal international trade regime has yet to seriously falter. Protectionist inroads have certainly been made in consumer electronics, steel, and automobiles, joining illiberal trade in agriculture and textiles. Yet trade in telecommunications, aircraft, and services has been liberalized. The 1993 Uruguay Round agreement may be indicative of the maintenance, if not promotion of international interdependence in trade.

These two points are tied together in this study by seeking to explain the maintenance of liberal trade relations in terms of the mutual economic dependence of industrial countries. In particular, we will examine what may be a fundamental constraint on trade protectionism today: the reliance of industrialized economies on external trade relations, and especially on markets within the industrial world. This export dependence thesis is explored primarily for the case of the

United States, although comparative implications for other industrial countries will also be addressed.

The remainder of this chapter elaborates the export dependence thesis and sets the substantive and theoretical backdrop against which it is set. In particular, the export dependence thesis is set against prevailing international political economy explanations of trade policy based upon import penetration and hegemonic stability. The thesis is also contrasted with the extensive literature on interdependence, dependence, and dependency.

The investigation of the export dependence thesis is conducted at three levels of analysis (industrial, national, and international) and in two modes of analysis (aggregate data and case study). Chapter II provides a first test for the thesis in a comprehensive aggregate data analysis of 129 United States industrial sectors. Varying levels of export dependence across the 129 sectors are directly related to their respective protectionist behaviors. The thesis is examined in the context of a vast array of protectionist instruments including escape clause, antidumping, and countervailing duty motions, voluntary export restraints, orderly marketing arrangements, quotas, the steel trigger price mechanism, and the Multifiber Arrangement. Among the most significant findings of this analysis is that industries dependent upon exports manifest relatively low levels of protectionism *even if they are also highly penetrated by foreign imports*. Industries threatened by import penetration evidence protectionist activity only if they are also not dependent upon foreign markets for sales.

Chapters III and IV provide a further test of the export dependence thesis in a crucial case study of the U.S. machine tool industry. The machine tool case is "crucial" because the export dependence thesis is pitted against what would seem to be overwhelming pressures propelling the industry towards protection. Despite experiencing severe and ever-intensifying import penetration, dramatic hegemonic decline, cyclical economic performance, and diminishing if not dissipated technological superiority worldwide, the industry maintained relatively liberal trade policy views through the 1970s. This anomalous outcome can be directly attributed in part to the industry's high levels of export dependence.

Chapter V extends the analysis to other industrial countries and to international trade relations generally. Specific observations are made regarding differential levels of constraints on protectionism emanating from export dependence in North-North versus North-South trade relations. This final chapter concludes by drawing out the study's implications for general theory in international political economy, and particularly for the study of national power and the structure of the international political economy.

This study has been theoretically as opposed to substantively motivated. At each level and mode of analysis we have sought to examine how much one can explain with our sole hypothesis. As such, comprehensive explanations of U.S. trade policy or international trade relations are not sought nor is an all-encompassing examination of the U.S. machine tool industry conducted. While the export dependence thesis constitutes our theoretical contribution, no pretense is made to the effect that complex outcomes in business and government policy can be accounted for by monocausal explanations. Without losing sight that "other factors" are always at play, our purpose here is to examine the explanatory power of one such factor, export dependence.

THEORIES OF TRADE POLICY

The core of this study is the explanation of protectionism or, more properly, the explanation of its absence at the industry level. We seek to determine whether or not industries seek trade protection. The thesis does not address the issue of whether industry demands, once in the governmental realm, are met or not.

An explanation of industry behavior is a fundamental issue that may logically precede an explanation of governmental response. In industrial countries, industry sources constitute a major breeding ground for protectionism. Government actors are often only secondary sources of protection. If and when official protection is adopted, government action is often (reluctantly) taken in reaction to industry demands. The state may shape, rechannel, or deny such demands, but if protection is adopted (in whatever form), it is usually initiated by industry not government.

A great deal of analysis has been devoted to how governments formulate foreign economic policy, often in the context of industry demands.[3] This study establishes that context. Instead of examining how industry trade demands get through the sieve of government machinery, we seek to determine which industry demands do or do not get to the governmental sieve in the first place. In sum, this thesis is directly and primarily concerned with the explanation of *industry* trade policy *views* and only indirectly and secondarily with *state* foreign trade *policy.*

In constructing this explanation we can, at the outset, set aside most of the general macroeconomic arguments against protectionism and for free trade.[4] We can do so because at the sectoral level, the economic fortunes of a protected industry may not be affected or may flourish while the economy as a whole suffers the costs of the protection gained by the one industry. First, while an economy as a whole suffers from the production costs of protection in terms of impaired specialization, resource allocation and efficiency, the protected industry can experience production gains. Second, the economy-wide consumption costs of protection in terms of higher prices, lower quality and reduced demand may not be considered instrumental to the protected industry. As one analyst of the production and consumption costs of protection has observed:

> On the micro-economic level, decision makers cannot be expected to take into account those benefits and costs generated by their operations which are diffused over the entire economy.[5]

Even one of the most visible costs of protection — foreign retaliation — may not be relevant to the protected industry. Even if foreign retaliation is undertaken, the industry whose protection provoked the retaliation may not be affected. For instance, in response to the raising of U.S. duties on sheet glass in 1962, the EEC suspended tariff concessions on polyethylene, synthetic fibers, and varnishes.[6] In the "chicken war" of 1963, the United States reversed tariff reductions on starches, brandy, and trucks in retaliation against EEC levies on poultry imports.[7] Clearly, the macroeconomic costs of protection need not act as effective deterrents against protectionism at the sectoral level.

The Export Dependence Thesis

Once a sectoral level of analysis is adopted, however, *and if an industry is dependent upon exports*, then protectionist actions do indeed entail prohibitive costs, especially export price and foreign supply costs.

Export price costs. The first primary cost of protection for an export dependent industry concerns price competitiveness. Protection by any means raises the cost of the restricted imported good. Irrespective of the encouragement of industrial inefficiencies, protection induces price increases for domestically-produced substitutes through the market mechanism. And just as domestic prices may be raised, the prices of exportables directly competitive with the foreign goods are pushed up.[8] Such price increases undermine the international competitiveness of exportables.

The price costs of protection are heightened under conditions where trade in intermediate goods is significant and where manufacturers are diversified vertically. Trade in intermediate goods — semimanufactures (e.g., yarn, cement, steel) used as inputs for other semimanufactures or final goods — constitutes 40% each of OECD exports and imports in manufactures.[9] This substantial trade has widespread ramifications for foreign economic policy.[10] Of interest to this study are the effects of protection on intermediate goods for exports. If foreign intermediate goods are necessary for domestic production, protection against such inputs will raise the costs of production. The same increase in the costs of production and, therefore, in final price applies as well to exportables that utilize foreign inputs.[11]

Clearly users of intermediate products have motivations to oppose protection on such goods. Yet given high and increasing levels of vertical integration in manufacturing, firms have become producers as well as users of intermediate goods.[12] These intermediate goods producers, in the interests of their final goods producing branches, may also oppose protection.

Foreign supply costs. The second primary cost of protection concerns foreign supply developments that adversely affect the international markets of export dependent industries. To begin with, a foreign supplier, finding access to a protected market restricted, may channel more of its shipments towards its own domestic market.[13] The ability and effects of an industry resorting to greater domestic sales in response to protection abroad are enhanced under conditions of intra-industry trade. Intra-industry trade, or the simultaneous importing and exporting of similar if not identical products, has been shown to be significant and increasing in transactions between industrial countries.[14] If a participant in such trade finds its foreign markets restricted or denied, it can redirect its products to its own well-developed domestic market. By increasing competition in the targeted country's markets, protection endangers an industry's export position in those markets.

Even when intra-industry trade conditions do not exist and the market of the targeted country is not significant for the export dependent industry, protection for that industry may threaten its export positions in third country markets. Having been denied access to one market, a foreign supplier can direct its products to other foreign markets. As a direct result of protecting its home market, an industry may face greater competition from the restricted foreign suppliers in third country markets.

Such dynamism in international trade flows has long been demonstrated especially in reaction to trade barriers. For instance, since the 1970s, the U.S. and Europeans have each feared that restrictions on Japanese steel, consumer electronics, and automobiles on one side of the Atlantic would lead to further market penetration by Japan on the other (their) side of the Atlantic. A similar shift in the destination of export shipments, in response to restrictions in one market, can threaten not only the domestic markets of other countries, but also the export markets of the protected industry.

These price and supply costs of protection act to directly inhibit protectionist actions on the part of export dependent industries. Given the importance of export sales to such industries, these two costs are unacceptable. The loss of such sales, like any other sizable market loss, would threaten an industry's stability and growth. In times of domestic recession — of contracting home markets and progressive underutilization of industrial capacity — a firm's foreign sales become all the more vital.

Secondary Costs. The export price and foreign supply costs of protection serve as the major constraints against the resort to protectionism by export dependent industries. Two other, less instrumental or less direct costs could have similar effects. One of the secondary constraints may be labeled the "infrastructural" costs of protection. An industry dependent upon the international arena may resist protectionist pressures within its own ranks and in other industries. Any deviation from free trade practices may be viewed as contributing to the long-term systemic deterioration of the international trade regime. Here the concern is not so much with the immediate repercussions of protection, but with the weakening of liberal trade infrastructure.

The other secondary constraint concerns the costs from potential foreign retaliation. We have already noted that foreign retaliation may not affect the industry that gains protection. Yet given the fact that they will be heavily engaged in foreign markets, export dependent industries become prime targets for retaliation. Such industries are susceptible to and would seek to avoid the raising of trade barriers against their products.

These costs of protection, especially the export price and foreign supply costs, are significant not because they apply to other sectors of the economy (they do this also), but because they apply to the very industry that gains protection.[15] Unlike the case for the macroeconomic costs of protection, export dependent industries directly pay the price for protectionist indulgence.

We began this section by stressing the need to explain trade policy at the *industry* level. Yet that same rationale establishes the relevance of the export dependence thesis for *national* trade policy. If industry behavior is a major input for state policy, then constraints imposed by export dependence on industrial trade policy translate into constraints on the formulation of national foreign economic policy. Industrial opposition to certain policy options heightens the restraints on national policy even if such opposition is not completely effective. This linkage to national trade policy is buttressed where large numbers of industries are subject to the constraints of export dependence (as will be found in the U.S. case in Chapter II).

Import Penetration and Hegemonic Stability

Proposing that export dependence has a role to play in constraining protectionism goes against the grain of much of the literature on the future and stability of trade relations. In the first place, much of this literature ignores the importance of export trade and, instead, focuses on the problems posed by imports, especially the import penetration of domestic markets.[16] An often-cited source of protectionism is the growth in post-war economic interdependence and, particularly, the growth of its adversities for domestic economies. Countries have become more exposed to external economic disruptions, particularly import surges that threaten businesses, jobs, and communities. Protectionism is fueled by demands to diminish such national vulnerability.[17]

Besides the growth of international interdependence, three other developments have exacerbated the disruptiveness of imports and, thereby, pressures for protection: changes in comparative advantage, domestic economic recession, and the accumulation of surplus industrial capacity. Structural changes in comparative advantage across nations have led to alterations in global supply and demand patterns. The postwar dynamism of such changes have left in their wake "declining industries" in many industrial countries, particularly in textiles, steel, and shipbuilding.[18] The most visible manifestation of shifts in comparative advantage is the rise of greater volumes and more numerous suppliers of imports into countries once efficient producers of the imported products.

Under high-growth conditions, the costs of structural adjustment are reduced by sacrificing extra-production increases. Conversely, periods of economic recession make structural adjustments exceedingly difficult and costly.[19] In general, periods of recession and unemployment become "breeding grounds" for protectionism;[20] the 1973-74 worldwide recession provided the major impetus for the emergence of the new protectionism. Where depressed industries face foreign competition, imports become easy scapegoats for economic ills and protection a "cheap" policy for "recovery."

The end result of domestic resistance to structural adjustment when confronted with changes in comparative advantage is the accumulation of surplus, cost-inefficient industrial capacity.[21] Excess capacities in industrialized countries are all the more vulnerable to the imports of newer, more efficient producers; a formula for protection.

Related to import penetration,[22] the other major focus of theories of protectionism is the distribution of power internationally. Liberal trade relations are considered possible only under the aegis of a hegemonic power (e.g., Britain in the mid-nineteenth century and the U.S. after the Second World War).[23] Economic and technological superiority worldwide predisposes a hegemon towards a liberal international economic regime. It constructs and maintains such a regime, first, by providing economic incentives for the participation of other countries. Such incentives entail the provision of collective goods: lender of last resort, market for distress goods, supplier of capital and financial services. Second, the hegemon gains the compliance of other countries and limits protectionist deviations with its coercive power — notably, its ability to link international economic issues to other issues such as the provision of military security. For theorists of hegemonic stability, the relative decline in U.S. economic and political dominance over the past two decades provides yet another reason to anticipate the dismantling of the liberal trade regime.

The purpose of this study is not to dismiss these undeniably potent pressures for protection. Instead, the aim is to offer and test export dependence as an imperative with the potential to offset protectionist tensions. Among the implications of this research is the recognition that import penetration need not be a sufficient condition for protectionism nor hegemony a necessary requisite for liberal trade.

INTERDEPENDENCE, DEPENDENCE, AND DEPENDENCY

The export dependence thesis builds upon but is distinct from much of the literature on interdependence, dependence, and dependency. Clear, although frequently misconstrued distinctions have been drawn between: (a) sensitivity interdependence - the sensitivity of economies to international transactions or other external forces; (b) dependence - an economy's reliance on external economic relations (or what has also been called "vulnerability interdependence"); and (c) dependency - the external control and structural distortion of an economy.[24] As used in this study, "dependence" is used in the second sense only.

Interdependence refers to the permeability of a domestic economy to external interventions. The concern is with the openness or susceptibility of an economy to fluctuations in trade, monetary or capital flows, and to the policies of other countries affecting such flows. At issue is the breaking down of bounded national economic units. In contrast, the concept of dependence assumes the porousness of such boundaries and shifts the focus to the adaptation and transformation of internal economic structure. The issue here is the extent to which an economy is structurally oriented towards and relies upon indigenous versus foreign markets for accumulation, production and consumption.[25]

The effects of hypersensitivity to external market forces (interdependence) and of reliance on external sources for the completion of basic economic activities (dependence) underlie the conditions embodied in the concept of dependency: the malintegration or disarticulation of economic structure and lack of self-sustaining economic growth.[26] In a similar fashion, conditions of interdependence and dependence provide two contexts under which participation in the world economy limits foreign economic policy options on such issues as trade barriers, openness to foreign capital, and exchange rate policy.

The following two sections establish the distinctiveness of the export dependence thesis from much of the work on interdependence, dependence, and dependency. In contrast to the extensive literature on the adversities of international interdependence, the constraints of export dependence are based on fundamental relations of dependence rather than mere sensitivity to external disruption. Similarly, export dependence constitutes a structural as opposed to a process or behavioral constraint on policy. Finally, and at variance with most of the international political economy literature, export dependence acts as an internalized rather than a purely external policy constraint.

Dependence versus Interdependence

Analysts of *sensitivity interdependence* have examined how economic developments that originate outside of the national arena impact on domestic as well as foreign economic policy.[27] The analysts are concerned with:

> The increasing sensitivity of economic events in one country to what is happening in its trading partners. (Cooper, 1968, p. 10)
>
> The problem of controlling events external to the state. (Morse, 1970, p. 373)
>
> Transnational forces impinging on [nations] from outside their boundaries. (Keohane and Nye, 1972, p. 394)
>
> The economic responses of one country to another. (Rosecrance, et al., 1977, p. 429)
>
> Foreign exogenous influences [playing] an enormously important part in influencing many macrovariables in the home economy. (Bryant, 1980, p. 138)

In each case, the concern is with the external influence emanating from an economy's interdependence with the rest of the world and not with reliance on external relations.

Trade disruptions are among the most basic illustrations of the adversities of interdependence. To take one example, imports of color television in the mid-1970s have been held responsible for severe reductions in the U.S. television industry in terms of the number of operating firms, capacity utilization, and employment.[28] The impact of international interdependence is also epitomized by the volatility of short-term financial flows. The classic case is the West German experience in May 1969. Financial flows from external sources inflated the country's money supply by 20% at a time when the Bundesbank was pursuing contractionary policies.[29]

Trade and financial disruptions evidence the effects of the interdependence of international markets. The interstate equivalent is the interdependence of national policies. Interstate policy sensitivity is illustrated by the exchange rate manipulations of the interwar period.[30] Whether or not North American and European exchange-rate policies were purposefully "competitive" (i.e., each designed to undercut the other), their effect was to undermine each other's effectiveness in gaining price advantages in export markets.[31]

A purer case of competitive national policies has been the rivalry among industrial countries over export credits. Policy interdependence is demonstrated in this case not just by efforts to undercut trade competitors by offering ever-more-concessional terms to importers, but by the multinational negotiations aimed at unraveling the vicious cycle.[32]

Countries can be sensitive to the economic behavior of transnational actors as well as state actors or international market forces. For instance, dependency as well as non-dependency studies have shown how, in pursuing their global economic interests, transnational enterprises (TNEs) can adversely affect economic policy and economic development in developing countries. For instance, the entrance of a multinational can entail the acquisition of local firms and the absorption of local capital with dire consequences for indigenous industrial development efforts.[33]

Foreign subsidiaries may prefer to obtain supplies from their parent company rather than local suppliers. Such an import orientation could not only damage the host country's balance of trade, but could severely limit the TNE's backward and forward linkage for developing indigenous manufacturing and service sectors. The host economy's balance of trade could also be adversely affected by the manipulation of transfer prices in intrafirm trade where subsidiary exports are underpriced and imports overpriced.[34]

The economic impact of TNEs on a host country's public policy is underlined by Theodore Moran's observation regarding the Chilean experience:

> No domestic decision about the rate of economic development, the strength of the balance of payments, the level of aggregate employment, the breadth of social welfare programs could be made without a careful consideration, from outside, of how

> the foreign copper companies might be going to exercise their discretionary power.[35]

The sensitivity of countries to outside forces serves as the basis for the policy constraints discussed thus far. Analysts of economic power, however, have explored a very different foundation for international influence: relations of *dependence.*[36]

For these analysts, one country's dependence upon another yields a resource for power and influence for the other country:

> The influence which country A acquired in country B by foreign trade depends in the first place upon the total gain which B derives from that trade; the total gain from trade for any country is indeed nothing but another expression for the total impoverishment which could be inflicted upon it by the stoppage of trade. In this sense the classical concept, gain from trade, and the power concept, dependence on trade, now being studied are seen to be merely two aspects of the same phenomenon. (Hirschman, 1945, p. 18)

> Power resides implicitly in the other's dependency. . . the dependence of one party provides the basis of power for the other. (Emerson, 1962, p. 32)

> It is asymmetries in dependence that are most likely to provide sources of influence for actors in their dealings with one another . (Keohane and Nye, pp. 10-11)

> Strategic advantages [are] to be had by creating and nurturing economic dependencies. (Richardson, 1978, p. 66)

> International economic leverage . . . is inherent in the symmetry of mutual dependencies. (Knorr, 1979, p. 168)

As noted by Albert Hirschman, it is the high cost of having relations of dependence broken which makes dependence the "root cause" of influence for others.[37] The policy constraint from interstate dependence can be exemplified by two historical episodes: Germany's economic domination over Southeastern Europe in the 1930s and the "Sterling-Dollar diplomacy"[38] of the 1940s.

The regional importance of Hitler's Germany as supplier of manufactures and market for agricultural goods provided the Nazi regime a great deal of leverage over the countries of Southern and Eastern Europe. Bilateral commercial agreements were exacted from Germany's satellites establishing near monopoly over their imports and near monopsony over their exports. By inducing agricultural exports to Germany, the commercial agreements were designed to throttle industrialization in Southeastern Europe.[39]

German regional economic domination was reinforced by bilateral clearing arrangements whereby trade payments were bypassed with a system of claims on each party's goods. Often Germany would intentionally run balance of trade deficits with its satellites where the unspent claims represented forced loans to Germany. Since claims could be used to purchase only German goods, the satellites directed their foreign purchases toward Germany even though prices for goods may have been lower elsewhere.[40]

By the outbreak of the Second World War, the U.S. was committed to constructing a multilateral economic order based on reduced and nondiscriminatory trade and payments restrictions. Great Britain shared U.S. objectives for postwar international collaboration and tariff reduction. Britain balked, however, on the principle of non-discrimination since her special arrangements with the Commonwealth in trade (Imperial Preference System) and payments (Sterling Area) were considered necessary for postwar economic recovery. The same war devastation that heightened Britain's need for these discriminatory arrangements, however, also meant that the country's financial dependence on the U.S. during the war would extend into the postwar period as well. The U.S. would use that dependence to gain the dismantling of Britain's discriminatory trade and payments practices.[41]

The U.S. first exercised its financial leverage in reaching the 1942 Mutual Aid Agreement that governed lend-lease aid to Britain. In exchange for the aid, the U.S. gained a reluctant British commitment to cooperate in efforts to realize America's postwar liberal international economic order.[42]

With the termination of lend-lease at war's end, British dependence upon the U.S. heightened and the U.S. exercised its financial leverage in negotiating the 1945 British loan. Britain's obligations under the $3.75 million loan agreement did not end (or really even begin) with the 2% per annum interest charge. To further its postwar international economic objectives, the U.S. exacted from

Britain: (a) a commitment to make sterling convertible within one year; (b) the elimination of exchange and quantitative trade controls that discriminated against U.S. goods; and (c) the scaling down of sterling reserves held in Sterling Area countries that could be used by Britain to gain markets at the expense of U.S. exports.[43]

Clearly, the policy constraints imposed in these two cases are based on fundamental relations of dependence and not merely on sensitivity to external forces.

Countries may become dependent not just on other countries, but on nonstate actors as well. Among the most visible manifestations of policy constraint emanating from dependence on nonstate actors are the stabilization programs required of less developed countries (LDCs) seeking loans from the International Monetary Fund. The conditions for receiving IMF loans are profound for public policy and are almost "stock formula":[44] liberalization or elimination of currency and import controls, exchange rate devaluation, limits on domestic credit expansion, reductions in public deficit and spending, increased taxes and prices, cuts in public subsidies, reductions in state enterprise holdings, wage restraint, encouragement of foreign investment. The ability of the IMF to impress such conditions rests not just with the provision of needed credit. More importantly, an agreement with the IMF represents a stamp of approval for access to the much larger resources of commercial banks.

In fact, by 1978 57% of LDC debt was with private sources,[45] and six U.S. banks held approximately two-thirds of those private claims.[46] With such substantial resources concentrated in so few hands, major international banks wield influence akin to that of the IMF. For example, in July 1976 a consortium of U.S. banks headed by Citibank and including the Bank of America, Chase Manhattan, Manufacturers Hanover, Morgan Guaranty, and Wells Fargo reached a $400 million loan agreement with Peru that included IMF-like conditions: a 44% devaluation, price increases, budget cuts; more favorable treatment of foreign investment, including reopening areas to private oil companies and reaching compensation and taxation agreements with specific foreign firms; and changes in labor legislation to attract more private investment. The agreement provided for the unprecedented monitoring of the Peruvian economy by private banks.[47]

Beyond sensitivity to the microeconomic behavior of transnational enterprises, dependence on TNEs for capital and technology can entail macroeconomic policy constraints. State-TNE negotiations are the most explicit arenas where such dependence can compromise national policy in such areas as equity control, employment and labor policies, research and development priorities, import and export policies, taxation, forward and backward linkages, and reinvestment.[48]

Structural versus Process Constraints

The distinction between constraints based upon dependence and interdependence corresponds roughly to that between *structure* and *process*. International processes encompass the short-term behavior of actors. Actors can be affected by or affect others through such process interactions as international negotiation or international coercion. International structure, on the other hand, concerns long-term situational factors that determine the incentives and constraints within which actors operate. In this case, we refer to how actors can be influenced not by their discrete interactions, but by such contextual pressures as the distribution of international power or the international division of labor. In these cases, lessor actors may be deterred from acting or compelled to act, and greater powers may have long-term control over events without the physical exertion of power or even the threat thereof by the one over the other.[49]

Sensitivity interdependence provides a context for international process influence. The constraints that emanate from international economic disruptions, the policies of foreign countries, or the economic behavior of transnational actors are all based upon concrete interactions of an actor with the rest of the international system.

In contrast, economic dependence or the reliance on external economic relations provides a context for international structural influence. The ability of the U.S. to exact concessions from Britain during and immediately after the Second World War was based less on U.S. bargaining tactics than on a highly asymmetrical relationship between a debtor and creditor.

External versus Internalized Constraints

Structural power often affords wide latitude for the exercise of process or behavioral power. Highly asymmetrical structural relations permitted Nazi Germany and the U.S, to directly impose their will on Eastern Europe and Britain, respectively. Yet the most interesting "face"[50] of structural influence is that its "effects" need not be imposed by tangible, observable means. In recognition of structural limitations, actors may unilaterally refrain from certain policies or adopt others. The end result, some form and extent of behavioral compliance, may differ little from outcomes explicitly imposed by others.

We are then led to another important distinction (formally in types of structural constraint) that helps to further illuminate the export dependence thesis. All the international economic constraints discussed in the prior section dealt with restraints imposed by *external* forces. The "effects" of sensitivity interdependence emanate from abroad whether they be external markets, state actors, or non-state actors. Vulnerability to the machinations of external actors, state or non-state, emanates from highly asymmetrical relations of dependence. Indeed, a vast proportion of the international political economy literature concentrates on the imposition of external, international constraints on the formulation of national policy. This emphasis is natural since much of the literature derives from the more general study of international relations, but this bias should be explicitly recognized.

In contrast, an important feature of the present argument is that constraints on foreign policy need not be imposed from abroad. Beyond purely domestic constraints (see below), the narrowing of policy options can be *internalized* in light of relationships with the international arena.[51] Such internalization or unilateral adjustment to structural limitations can best be illustrated by dependence on (and not interdependence with) the international market (rather than dependence on specific actors). The constraints of market dependence encompass the export dependence thesis.

Dependence on the global market economy for supplies (merchandise, capital, technology) and markets can place formidable demands and restrictions on national policy. For instance if, as is the case for many LDCs, a country is dependent on imports of consumer and/or capital goods, there are strong pressures to promote exports to gain foreign exchange. The export requirements of import dependence can frustrate national efforts to diversify exports (by requiring concentration on proven export products) and to develop domestically-oriented industry and domestic market demand (by shifting resources to export industries and tending to foreign demand). To the extent that their export revenues fall far short of import requirements,[52] developing countries have been forced to turn to foreign capital or international borrowing to "alleviate" the problems of import dependence.

We have already discussed the external constraints that can be imposed by transnational enterprises and international banks. Yet dependence on foreign capital or international loans as general economic conditions can entail self-imposed policy requirements. For instance, to establish a "good investment climate" for sorely needed foreign capital, Chile adopted new mining legislation in 1955 that included reduced tax rates, the elimination of artificial exchange rates, accelerated depreciation rates for new investment, liberal expense allowances for mining development, free import rights on equipment, and the dissolution of the state monopoly over copper sales.[53] Similarly, in 1964 the Brazilian military regime abolished subsidies for domestic business and reduced tariff levels to attract international capital.[54]

Debt dependence can also entail the internalization of policy constraints. Before Peru made loan requests to major international banks in 1976, the government imposed a stabilization package including a wage freeze, devaluation, and public sector budget cuts; declared a state of emergency canceling union rights to strike; ended the ban on new oil contracts with foreign firms; and agreed to pay compensation for the expropriation of a foreign-based mining company.[55]

To improve its weakening credit standing, in 1971 the Brazilian government dismantled its state monopoly in oil exploration to allow participation by foreign companies. the policy reversal was taken "to show the international financial community that Brazil welcomed foreign currency, loans, and capital investment.[56] In November 1980, Brazil eliminated price and rent controls, and scrapped its managed devaluation and plans for two-tiered interest rates when the planning minister recognized that he had "more or less exhausted the [international] banks' willingness to lend his country money — at least until Brazil is back in the black."[57]

Similarly, in July 1981 India undertook a series of self-initiated economic measures four months in advance of its IMF loan request. Domestic oil prices were trebled, fertilizer and food subsidies slashed, domestic credit expansion tightened, the budget deficit cut, and import controls liberalized despite India's huge trade gap ($6.8 billion deficit in fiscal year 1980).[58]

The central characteristic in each of these cases is that none of the policy measures undertaken were dictated from abroad. While some may have been taken in anticipation of foreign demands or reactions, all the policy actions reflected, in a fundamental sense, national economic positions vis-à-vis the rest of the world. The self-imposed measures recognized the policy imperatives that directly flow from dependence on foreign imports, capital, or finance.

In a similar fashion, export dependence establishes internalized not external constraints on protectionism. As discussed earlier, prohibitive export price and foreign supply costs of protection are inherent under conditions of export dependence. These two costs are not so much based on the actions of others as they are built-in consequences of export dependence. Just as conditions of import, capital or debt dependence can independently compel countries toward certain policies, export dependence discourages protectionism. In each case, national policy freedom is compromised in order to seek needed benefits from the international economy and not as a result of external coercion. The constraints of export dependence differ from the other forms of market dependence only in dealing with the implications of reliance on international markets rather than foreign supplies.

If the constraints of export dependence and international market dependence in general are not external ones, they also are not purely domestic pressures (e.g., the impact of interest groups, political organizations, social coalitions, state structures). While the internalized constraints of export dependence are domestically imposed, it must not be forgotten that they reflect the imperatives of vulnerable positions in relationships with the rest of the world. While economic dependence is a domestic condition, its sources are both the structural orientation of the domestic political economy as well as participation in the global economy. The constraints of export dependence highlight the disintegration of the boundary between domestic and international affairs.

INDUSTRIAL COUNTRIES, MANUFACTURED EXPORTS, AND THE U.S. CASE

The first part of this chapter established the rationale for seeking an explanation of trade policy at the industry level. In this section, we set out the basis for the specific empirical test for our explanation. In particular, we justify the focus on dependence on manufactured exports rather than some other form of international economic dependence, and on industrial countries and particularly the United States as opposed to other countries.[59]

In the first instance, we look to features of international trade relationships to find constraints on protectionism. International financial relationships, the networks of transnational capital, and many other non-trade global economic processes can certainly contribute to the maintenance of a liberal trade regime. Yet these various issue areas may be distinct enough and the issue linkages between them limited to the extent that regime changes in each may be significantly independent.[60] Indeed it would be possible for international financial and capital relations to become more closely interdependent while commercial relations grow more protectionistic. In light of such a possibility, it would be most appropriate to seek an intra-issue area explanation of liberal or illiberal trade. Hence of all the many transactions upon which an economy may be dependent, this study focuses only on reliance on external trade relations.

Secondly, we examine only trade in manufactures because such trade constitutes 60% of United States and 66% of OECD trade.[61] Another reason to study trade in manufactures is that the U.S. share of world exports in manufactures has steadily declined from 21% in 1955 to less than 12% in 1978; a 45% decline (see Figure 1.1). Such a deterioration in U.S. dominance would be cited as a key reason to doubt the survivability of liberal trade relations (see above discussion of the theory of hegemonic stability).

Figure 1.1

U.S. Position in World Trade in Manufactures: Share of World Exports* and Import Penetration#, 1955-1980

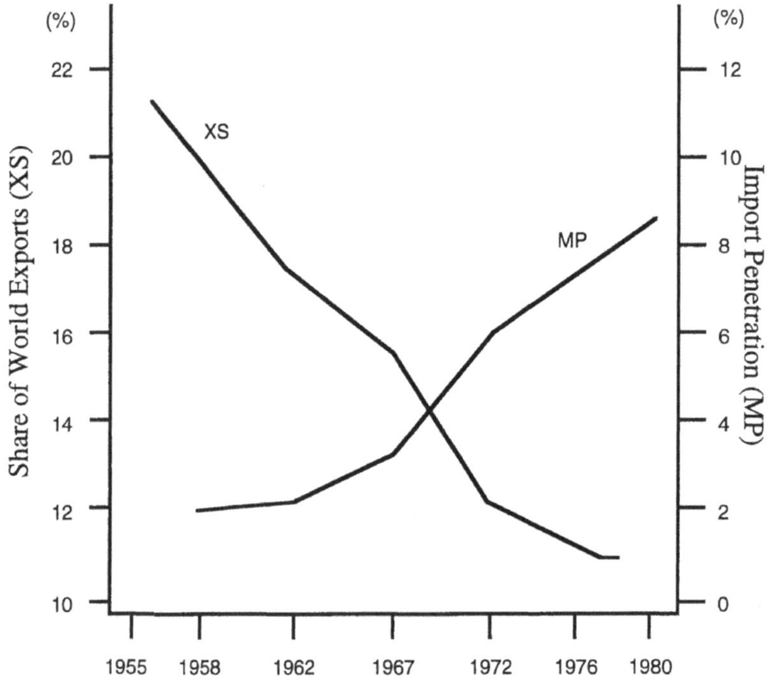

* SITC 5-8

\# ratio of imports to shipments plus imports minus exports; SIC 2-3

(continued)

Figure 1.1 (continued)

Sources: Export Share -
OECD, *Statistics of Foreign Trade, Annual: Tables by Reporting Countries, Series B*
U.N., Statistical Office, *Yearbook of International Trade Statistics*
UNCTAD, *Handbook of International Trade Statistics*
Import Penetration -
U.S., Bureau of Economic Analysis, *Survey of Current Business*
U.S., Bureau of the Census, *Annual Survey of Manufactures: Statistics for Industry Groups and Industries*
____, *Census of Manufactures*
____, *U.S. Commodity Exports and Imports as Related to Output,* ES2
____, *U.S. Exports: SIC Division by SIC-Based Two-Digits, Three-Digit, and Four-Digit Product Code,* EA 676
____, *U.S. Imports: SIC Division by SIC-Based Two-Digit, Three-Digit, and Four Digit Product Code,* IA 275

Third, we examine producer dependence on international trade rather than consumer dependencies. Due to better information and organization, producer interests generally have a greater impact on commercial policy than those of consumers.[62] Hence in exploring how economic dependence can constrain protectionism, it would be most appropriate to examine dependencies on the part of producers rather than consumers.

Fourth, we examine producer dependence on exports rather than imports. Dependence on (manufactured) exports is analyzed here because, in general, the potential for producer dependence on the outside world would be with exports and not, as would be the case for consumer dependence, with imports.[63]

Fifth, we focus on industrial countries rather than less developed countries. This orientation flows directly from the nature of the export dependence thesis. One of the critical assumptions of the thesis concerns the market size of the country or product line considered for protection. In order to incur many of the costs of protection, particularly the foreign supply, infrastructural, and retaliatory costs, the market to be closed must be relatively large. If the protected market is small, then the rechanneling of foreign goods and the heightening of competition in international markets will be minimal. Likewise, the closure of a small market may neither inflict damage to free trade infrastructure nor incite foreign retaliation. Countries with small markets (by product line or as a whole) may be able to "free ride" with protection at home without incurring costs nor forsaking the benefits of participating in a liberal trading regime.

Hence, in order for export dependent countries and industries to fear the costs of protection, their domestic market must be large enough so that its closure would be more likely to affect trade flows, weaken the trade regime, and incur foreign countermeasures. It is for these reasons that we focus on industrial countries. The export dependence thesis applies primarily to developed rather than developing countries since the former comprise the major markets for manufactures. Industrial countries are less able to "free ride" with protectionist policies; the ramifications of doing so are just too profound.

Finally, regarding the selection of the U.S. for case study, the United States represents a difficult and, therefore, theoretically significant case for testing the export dependence thesis.[64] First, the United States is characterized by a relatively "weak" state.[65] Separation of powers, federalism, bicameralism, the congressional committee system, judicial independence, autonomous bureaus, and "irresponsible" political parties contrast the U.S. political system with the more centralized state structures in most other industrialized countries: the centralized state apparatus with its extensive field system of administration in France; the centralized state bureaucracy in Japan; the corporatist structures in Scandinavia, Benelux, Austria, and Switzerland.[66] Given the fragmentation of the U.S. state, if so motivated, private sector protectionism could have the greatest impact on United States foreign economic policy. As Samuel Huntington has observed

> A strong state will not necessarily follow more liberal and internationalist economic policies, but a weak government is almost certain to be incapable of doing so.[67]

Second, the troubled U.S. economy has strong motivations to seek protection. As referred to earlier, the global position of U.S. manufacturing industries has suffered hegemonic decline. As a consequence, the U.S. has come to face increasingly greater market penetration by foreign manufactures (see Figure 1.1). Between 1958 and 1980, the share of the U.S. market for manufactures captured by foreign suppliers rose from 2.4% to 8.5%; a 250% increase. The challenge of import penetration plus hegemonic decline spells a potent formula for protection.

Third, in addition to having the motivation and "permissive conditions" for seeking protection, the constraints of export dependence would seem to be relatively low for the United States. The aggregate levels of manufactured export dependence in the industrial world are presented in Table 1.1. The United States is the least dependent industrialized country, exporting 8.6% of its production. In contrast, Sweden manifests the highest level of export dependence with 33.7% followed by the Netherlands with 28.9%.

Table 1.1

Industrial Country# Export Dependence* in Manufactures, 1977@

Austria:	25.76%	Japan:	11.71%
Canada:	19.80	Netherlands:	28.90 (1976)
Denmark:	26.90	Norway:	23.96
France:	18.55	Sweden:	33.74
Germany (F.R.):	23.84	United Kingdom:	17.98
Ireland:	21.93(1975)	United States:	8.58
Italy:	26.46		

* ratio of exports (SITC 5-8) to shipment (ISIC 2-3)
\# complete data for Belgium, Luxembourg, and Switzerland not available
@ unless otherwise indicated

Sources: OECD, *Statistics of Foreign Trade, Annual: Tables by Reporting Countries*
U.N., Statistical Office, *Yearbook of Industrial Statistics*
_____, *Monthly Bulletin of Statistics*

If the export dependence thesis can be successfully applied to the United States despite all the unfavorable conditions, then the findings would take on all the greater precedence. It is exactly this challenge that is taken up in Chapter II.

CONCLUSION

This study examines a process by which imperatives of international economic relations are internalized to constrain industry trade policy options. In this regard, the export dependence thesis differs from much of the "interdependence" literature which emphasizes the imposition of external developments on foreign policy-making. The thesis builds upon those portions of the dependence and economic power literature that highlight the internalized ramifications of vulnerability rather than the exploitation of such vulnerability by external forces.

The thesis complements the literature on the state in foreign economic policy-making. Analysts have examined the process by which private interests influence and government actors make foreign policy. Scant analysis, however, has been conducted to explain industry behavior in the first place. This study provides one approach to such an explanation.

By highlighting a constraint against the adoption of trade protectionism, the export dependence thesis most directly confronts prevailing theories of protectionism. By citing either the ravages of import penetration or the decline of U.S. hegemony, mainstream analysts predict the rupture of liberal trade relations. This study will demonstrate that export dependence can act as a bulwark against protectionist pressures.

NOTES

1 Throughout this study, "industrial" countries, economies, markets, or world refers to Austria, Belgium, Canada, Denmark, France, Germany (Federal Republic), Ireland, Italy, Japan, Luxembourg, the Netherlands, Norway, Sweden, Switzerland, the United Kingdom, and the United States. This country-categorization is based on that of the International Monetary Fund.

Given the time period covered by this study, historically accurate references will be made to "West Germany," "Soviet Union," and various "socialist" countries of "Eastern Europe" without the modifier "former."

2 U.S., International Trade Administration, *International Economic Indicators*; U.S., President, *Annual Report of the President of the United States on the Trade Agreements Program.* See also Peter J. Katzenstein, "International Interdependence: Some Long-Term Trends and Recent Changes," *International Organization*, v. 29, n. 4 (Autumn 1975); Stephen D. Krasner, "State Power and the Structure of International Trade," *World Politics*, v. 28, n. 3 (April 1976); Clark A. Murdock, "Economic Factors as Objects of Security," in *Economic Issues and National Security*, Klaus Knorr and Frank N. Trager, eds. (Lawrence, Kansas: Regents Press of Kansas, 1977).

3 See Robert E. Baldwin, "The Political Economy of Postwar U.S. Trade Policy," *Bulletin, New York University Graduate School of Business*, n. 4, 1976; "Protectionist Pressures in the United States," *Challenges to a Liberal International Economic Order* (Washington: American Enterprise Institute, 1979); Raymond A. Bauer, Ithiel de Sola Pool, and Lewis Anthony Dexter, *American Business and Public Policy: The Politics of Foreign Trade* (Chicago: Aldine, Atherton, 1972); I.M. Destler, *Making Foreign Economic Policy* (Washington: Brookings, 1980); Judith L. Goldstein, "The State, Industrial Interests and Foreign Economic Policy: American Commercial Policy in the Postwar Period" (unpublished manuscript); Kent H. Hughes, *Trade, Taxes, and Transnationals* (New York: Praeger, 1979); Stephen D. Krasner, *Defending the National Interest: Raw Materials Investments and U.S. Foreign Policy* (Princeton: Princeton University Press, 1979); Timothy J. McKeown, "Hegemonic Stability Theory and Nineteenth Century Tariff Levels in Europe," *International Organization*, v. 37, n. 1 (Winter 1983); Robert A. Pastor, *Congress and the Politics of U.S. Foreign Economic Policy: 1929-1976* (Berkeley: University of California Press, 1980); J.J. Pincus, "Pressure Groups and the Pattern of Tariffs." *Journal of Political Economy*, v. 83, n. 4 (August 1978); Edward John Ray, "The Determinants of Tariff and Non-Tariff Trade Restrictions in the United States," *Journal of Political Economy*, v. 89, n. 1 (February 1981); E.E. Schattschneider, *Politics, Pressures and the Tariff: A Study of Free Enterprise in Pressure Politics as Shown in the 1929-30 Revision*

of the Tariff (Englewood Cliffs, NJ,: Prentice-Hall, 1935).

4 For comprehensive analyses of international trade theory as it relates to protection see W.M. Corden, *The Theory of Protection* (London: Oxford University Press, 1974); "The Costs and Consequences of Protection: A Survey of Empirical Work," in *International Trade and Finance: Frontiers for Research*, Peter B. Kenen, ed. (Cambridge: Cambridge University Press, 1975); Harry G. Johnson, *Aspects of the Theory of Tariffs* (Cambridge: Harvard University Press, 1971).

5 Peter G. Elkan, *The Meaning of Protection* (Wellington: N.Z. Institute of Economic Research, 1977). p. 47.

6 Gerald Curzon and Victoria Curzon, "The Management of Trade Relations in the GATT," in *International Economic Relations of the Western World, 1959-1971*, Volume 1, Andrew Shonfield, ed. (Oxford: Oxford University Press, 1976).

7 *Ibid.*, U.S., Tariff Commission, *Operation of the Trade Agreement Program, 1963-64*.

8 For theoretical discussion of these price dynamics, consult the literature cited in footnote no. 4. For empirical analysis of these dynamics, see, for instance, Morris E. Morkre and David G. Tarr, *The Effects of Restrictions on United States Imports: Five Case Studies and Theory*, Staff Report of the Bureau of Economics to the Federal Trade Commission (Washington: Government Printing Office, June 1980).

9 OECD, *Statistics of Foreign Trade*. Intermediate goods: SITC 51, 52, 53, 54, 58, 59, 60. 1979 figures.

10 For the effects of intermediate trade on exchange-rate devaluation, see D.J. Coppcock, "Devaluation When Exports Have an Import Content," *Manchester School*, v. 39, n. 4 (December 1971); Satya P. Das, "Traded Intermediate Products and the Theory of Devaluation," *International Economic Review*, v. 21, n. 2 (June 1930); Roon-Lum Shea, "Imported Inputs, Devaluation and Balance of Payments: A Keynesian Macro-Approach," *Southern Economic Journal*, v. 43, n. 2 (October 1976).

For its ramifications for "effective protection," see Bela Balassa, "Tariff Protection in Industrial Countries: An Evaluation," *Journal of Political Economy*, v. 73, n. 6 (December 1965); Corden, *The Theory of Protection*; *Effective Protection*, H.G. Grubel and H.G. Johnson, eds. (Geneva: GATT and Graduate Institute of International Studies, 1971); Johnson, *Aspects of the Theory of Tariffs*.

For its consequences for import-substitution industrialization strategies, see Corden, *The Theory of Protection*, p. 51; Bela Balassa and Associates, *The Structure of Protection in Developing Countries* (Baltimore: Johns Hopkins Press, 1971).

11 For general discussions on the effects of protection on intermediate inputs on exports, see Corden, *The Theory of Protection*, pp. 51, 116; *Trade Policy and Economic Welfare*, p. 76; Gottfried von Haberler, *The Theory of International Trade with its Application to Commercial Policy* (New York: Macmillan, 1936), pp. 235-236; Michael Michaely, *Theory of Commercial Policy: Trade and Protection* (Chicago: University of Chicago Press, 1977); William Penfield Travis, *The Theory of Trade and Protection* (Cambridge: Harvard University Press, 1964), p. 149.

For theoretical economic analyses of the effects, see Balassa, "Tariff Protection in Industrial Countries"; Ravundra Batra and Francisco R. Cases, "Traded and Nontraded Intermediate Inputs, Real Wages, and Resource Allocation," *Canadian Journal of Economics*, v. 7, n. 2 (May 1974); J. Clark Leith, "The Effects of Tariffs on Production, Consumption and Trade: A Revised Analysis," *American Economic Review*, v. 61, n. 1 (March 1971); Ronald L. McKinnon, "Intermediate Products and Differential Tariffs: A Generalization of Lerner's Symmetry Theorem," *Quarterly Journal of Economics*, v. 80, n. 4 (November 1966); Michael Michaely, "The Welfare Costs of Negative Value-Added," *Journal of International Economics*, v. 5, n. 3 (August 1975); Roy J. Ruffin, "Tariffs, Intermediate Goods, and Domestic Production," *American Economic Review*, v. 59, n. 3 (June 1969).

For empirical studies on the effects see Balassa and Associates, *The Structure of Protection in Developing Countries*; Elkan, *The Meaning of Protection*; James R. Williams, *The Canadian-United States Tariff and Canadian Industry: A Multisectoral Analysis* (Toronto: University of Toronto Press, 1979).

12 Ruth J. Maddigan, "The Measurement of Vertical Integration," *Review of Economics and Statistics*, v. 63, n.3 (August 1981); Irvin B. Tucker and Ronald P. Wilder, "Trends in Vertical Integration in the U.S. Manufacturing Sector," *Journal of Industrial Economics*, v. 26, n. 1 (September 1977).

13 H. Peter Gray, *International Trade, Investment and Payments* (Boston: Houghton Mifflin, 1979), p. 148.

14 Antonio Aquino, "Intra-Industry Trade and Inter-Industry Specialization as Concurrent Sources of International Trade in Manufactures," *Weltwirtschaftliches Archiv*, b. 114, h. 2 (1978); Herbert G. Grubel and P.J. Lloyd, *Intra-Industry Trade: The Theory and Measurement of International Trade in Differentiated Product* (New York: John Wiley & Sons, 1975), chapter 3; Helmut Hesse, "Hypotheses for the Expansion of Trade Between

Industrial Countries, 1953-1970," in *The International Division of Labor: Problems and Perspectives*, Herbert Giersch, ed. (Tubingen, 1974); Emilio Pagoulatos and Robert Sorensen, "Two-Way International Trade: An Econometric Analysis," *Weltwirtschaftliches Archiv*, b. 111, h. 3 (1975).

Aquino found that intra-industry trade represented 57% of U.S. trade in manufactures in 1972.

15 Lest there be any misunderstanding, the costs of protection are not the controlling factors in the explanation of trade policy presented here. These costs are operative in constraining protectionism only when an industry is export dependent.

16 Bahram Nowzad, *The Rise in Protectionism* (Washington: International Monetary Fund, 1978); Edward John Ray, "The Determinants of Tariff and Non-Tariff Trade Restrictions in the United States," *Journal of Political Economy*, v. 89, n. 1 (February 1981).

17 Klaus Knorr, "Economic Interdependence and National Security," in *Economic Issues and National Security*; Edward L. Morse, "The New Economic Nationalism and the Coordination of Economic Policies," in *The New Nationalism: Implications for Transatlantic Relations*, Werner Link and Werner J. Feld, eds. (New York: Pergamon Press, 1979).

18 Richard Blackhurst, Nicolas Marian & Jan Tumlir, *Trade Liberalization, Protectionism, and Interdependence*, GATT Studies in International Trade, n. 5 (Geneva: GATT, November 1977); Susan Strange, "The Management of Surplus Capacity: Or How Does Theory Stand Up to Protectionism 1970s Style?" *International Organization*, v. 3, n. 3 (Summer 1979).

19 Blackhurst, *et al.*, *Trade Liberalization, Protectionism, and Interdependence*; Gerald Curzon, "Neo-Protectionism, the MFA and the European Community," *World Economy*, v. 4, n. 3 (September 1981).

20 Morse, "The New Economic Nationalism;" Nowzad, *The Rise in Protectionism.*

21 Peter F. Cowhey and Edward Long, "Testing Theories of Regime Change: Hegemonic Decline or Surplus Capacity?" *International Organization*, v.37, n.2 (Spring 1983); Nowzad, *The Rise in Protectionism*; Strange, "The Management of Surplus Capacity;" Loukas Tsoukalis and Antonio da Silva Ferreira, "Management of Industrial Surplus Capacity in the European Community," *International Organization*, v.34, n.3 (Summer 1980).

[22] Import penetration is a product of a decline in economic power. Theorists of hegemonic stability have charted the decline of American economic dominance in terms of the growth of sectoral trade deficits. See Stephen D. Krasner, "U.S. Commercial and Monetary Policy: Unraveling the Paradox of External Strength and Internal Weakness," in *Between Power and Plenty: Foreign Economic Policies of Advanced States*, Peter J. Katzenstein, ed. (Madison: University of Wisconsin Press, 1978); "The Tokyo Round: Particularistic Interests and Prospects for Stability in the Global Trading System," *International Studies Quarterly*, v. 23, n. 4 (December 1979).

[23] Robert Gilpin, "The Politics of Transnational Economic Relations," in *Transnational Relations and World Politics*, Robert O. Keohane and Joseph S. Nye, eds. (Cambridge: Harvard University Press, 1972); *U.S. Power and the Multinational Corporation: The Political Economy of Foreign Direct Investment* (New York: Basic Books, 1975); "Economic Interdependence and National Security," in *Economic Issues and National Security* ; Charles P. Kindleberger, *The World in Depression, 1929-1939* (Berkeley: University of California Press, 1973); "Systems of International Economic Origin," in *Money and the Coming World Order*, David P. Calleo, ed. (New York: New York University Press, 1976); Stephen D. Krasner, "State Power and the Structure of World Trade," *World Politics*, v. 28, n. 3 (April 1976); "The Tokyo Round."

See also David P. Calleo, "Decline and Rebuilding of an International Economic System," in *Money and the Coming World Order*; Calleo and Benjamin M. Rowland, *America and the World Political Economy: Atlantic Dreams and National Realities* (Bloomington: Indiana University Press, 1973); Robert O. Keohane, "The Theory of Hegemonic Stability and Changes in International Economic Regimes, 1967-1977," in *Change in the International System*, Ole Holsti, *et al.*, eds. (Boulder: Westview Press, 1980); *Balance of Power or Hegemony: The Interwar Monetary System*, Benjamin M. Rowland, ed. (New York: New York University Press, 1976).

[24] For a discussion of these terms, see David A. Baldwin, "Interdependence and Power: A Conceptual Analysis," *International Organization*, v. 34, n. 4 (Autumn 1980); James A. Caporaso, "Dependence, Dependency, and Power in the Global System: A Structural and Behavioral Analysis," *International Organization*, v. 32, n. 1 (Winter 1978); Raymond D. Duvall, "Dependence and Dependencia Theory: Notes Toward Precision of Concept and Argument," *Ibid.*; Robert O. Keohane and Joseph S. Nye, "World Politics and the International Economic System" in *The Future of the International Economic Order: An Agenda for Research*, C. Fred Bergsten, ed. (Lexington Books, 1973); *Power and Interdependence*.

Baldwin offers solid grounds for using "dependence" to refer to external reliance and not external control or distortion.

As used in the international political economy literature, "vulnerability interdependence" has the same meaning as dependence: the reliance on external relations or, negatively, the opportunity costs of breaking such relationships. I have chosen "dependence" to refer to this reliance because the terms "vulnerability" and "interdependence" convolute any meaning of external reliance. First, vulnerability can convey susceptibility to external influence (or sensitivity interdependence) as well as external reliance. Holsti and Keohane and Nye have, for instance, used vulnerability in the first sense. Second, interdependence is an international condition characterizing the relations and interactions between countries. Reliance on external relations is, in contrast, a national or domestic condition characterizing the institutional and structural orientation of an economy (see subsequent discussion of "external" versus "internalized" constraints). Kal J. Holsti, "A New International Politics? Diplomacy in Complex Interdependence," *International Organization*, v. 32, n. 2 (Spring 1978); Robert O. Keohane and Joseph S. Nye, "Transitional Relations and World Politics: An Introduction," and "A Conclusion," in *Transnational Relations and World Politics*.

25 Since they have distinct meanings, the concepts should not be defined in terms of one another (e.g., dependence as asymmetrical interdependence; symmetrical dependence as interdependence). The adjectives "asymmetrical," "balanced," and "mutual," and their antonyms can, of course, be used to specify the basic concepts. In particular, "mutual economic dependence" is used in this study to refer to common conditions of dependence across countries.

26 Fernando Henrique Cardoso and Enzo Faletto, *Dependency and Development in Latin America* (Berkeley: University of California Press, 1979); Theotonia DosSantos, "The Structure of Dependence," *American Economic Review*, v. 60, n. 2 (May 1970); Peter Evans, *Dependent Development: The Alliance of Multinational, State, and Local Capital in Brazil* (Princeton: Princeton University Press, 1979); Tamas Szentes, *The Political Economy of Underdevelopment* (Budapest: Akademia: Kiado, 1971).

27 Ralph C. Bryant, *Money and Monetary Policy in Interdependent Nations* (Washington: Brookings Institution, 1980); Richard N. Cooper, *The Economics of Interdependence: Economic Policy in the Atlantic Community* (New York: McGraw-Hill, 1968); "Economic Interdependence and Foreign Policy in the Seventies," *World Politics*, v. 25, n. 2 (January 1972); Keohane and Nye, "Transnational Relations and World Politics;" Edward L. Morse, "The Transformation of Foreign Policies: Modernization, Interdependence, and Externalization," *World Politics*, v. 22, n. 3 (April 1970); "Transnational Economic Processes" in *Transnational Relations and World Politics*; *Modernization and the Transformation of International Relations* (New York: Free Press, 1976); Richard Rosecrance and Arthur Stein, "Interdependence: Myth or Reality?", *World Politics*, v. 26, n. 1 (October 1973); Rosecrance, *et. al.*, "Whither Interdependence?", *International Organization*, v. 31, n. 3

(Summer 1977).

28 U.S., International Trade Commission, *Television Receivers, Color and Monochrome, Assembled or Not Assembled, Finished or Not Finished, and Subassemblies Thereof: Report to the President on Investigation No. TA-201-19 Under Section 201(b) of the Trade Act of 1974* (USITC Publication No. 808, March 1977).

29 Lawrence Krause, "Private International Finance," in *Transnational Relations and World Politics*; Robert Solomon, *The International Monetary: An Insider's View* (New York: Harper & Row, 1977); Susan Strange, "International Monetary Relations," in *International Economic Relations of the Western World*, Volume 2.

30 Leland B. Yeager, *International Monetary Relations: Theory, History, and Policy* (New York: Harper & Row, 1976), pp. 328, 353-54, 356-65, 372.

31 Yeager argues that while some of the exchange rate movements may have been predatory, many were haphazardly determined. Yeager, *International Monetary Relations*, p. 374.

32 *Economist*, October 27, 1980; December 27, 1980; February 14, 1981; July 25, 1981; October 10, 1981.

33 For examples see Evans, *Dependent Development*, pp. 121-131; and Richard J. Barnet and Ronald E. Muller, *Global Reach: The Power of the Multinational Corporations* (New York: Simon & Schuster, 1974), pp. 153, 155.

34 *Ibid.*, pp. 158-159.

35 Theodore H. Moran, *Multinational Corporations and the Politics of Dependence: Copper in Chile* (Princeton: Princeton University Press, 1974), p. 6.

36 David A. Baldwin, "Power Analysis and World Politics: New Trends versus Old Tendencies," *World Politics*, v. 31, n. 2 (January 1979); "Interdependence and Power;" Peter M. Blau, *Exchange and Power in Social Life* (New York: Wiley, 1964); Richard M. Emerson, "Power-Dependence Relations," *American Sociological Review*, v. 27, n. 1 (February 1962); Albert O. Hirschman, *National Power and the Structure of Foreign Trade* (Berkeley: University of California Press, 1945); Robert O. Keohane and Joseph S. Nye, *Power and Interdependence: World Politics in Transition* (Boston: Little, Brown, 1977); Klaus Knorr, *The Power of Nations: The Political Economy of*

International Relations (New York: Basic Books, 1975); "International Economic Leverage and Its Uses," in *Economic Issues and National Security*; "National Power in an Economically Interdependent World," in *Power, Capabilities, Interdependence: Problems in the Study of International Influence*, Kjell Goldman and Gunnar Sjostedt, eds. (London: Sage Publications, 1979); Neil R. Richardson, *Foreign Policy and Economic Dependence* (Austin: University of Texas Press, 1978); Kenneth N. Waltz, "The Myth of National Interdependence" in *The International Corporation*, Charles P. Kindleberger, ed. (Cambridge: MIT Press, 1970); *Theory of International Politics* (Reading: Addison-Wesley, 1979).

[37] Hirschman, *National Power and the Structure of Foreign Trade*, p. 116; Baldwin, "Interdependence and Power," p. 500, Blau, *Exchange and Power in Social Life*, pp. 119-120; Keohane and Nye, *Power and Interdependence*, p. 13; Waltz, "The Myth of National Interdependence, p. 210.

[38] The term is from Richard N. Gardner's classic study, *Sterling-Dollar Diplomacy: Anglo-American Collaboration in the Reconstruction of Multilateral Trade* (New York: McGraw-Hill, 1969).

[39] Hirschman, *National Power and the Structure of Foreign Trade*, p. 36.

[40] Yeager, *International Monetary Relations*, p. 369.

[41] Gardner, *Sterling-Dollar Diplomacy*, pp. 13, 28, 36, 42, 178-79, 191.

[42] *Ibid.*, p. 54-62.

[43] *Ibid.*, pp. 204-206, 213-221; Fred L. Block, *The Origins of International Economic Order: A Study of United States International Monetary Policy from World War II to the Present* (Berkeley: University of California Press, 1977), pp. 62-68.

[44] Charles Lipson, "The IMF, Commercial Banks, and Third World Debts," in *Debt and the Less Developed Countries*, Jonathan Aronson, ed. (Boulder: Westview, 1979), p. 324.

[45] Jeff Frieden, "Third World Indebted Industrialization: International Finance and State Capitalism in Mexico, Brazil, Algeria, and South Korea," *International Organization*, v. 35, n. 3 (Summer 1981), p. 409.

[46] Jane D'Arista, "Private Overseas Lending: Too Far, Too Fast?" in *Debt and the Less Developed Countries*, p. 74.

[47] Barbara Stallings, "Peru and the U.S. Banks: Privatization of Financial Relations," in *Capitalism and the State in U.S.-Latin American Relations*, Richard R. Fagen, ed. (Stanford: Stanford University Press, 1979), p. 239.

[48] For empirical cases, see Evans, *Dependent Development*; Moran, *Multinational Corporations and the Politics of Dependence*; Alfred Stepan, *The State and Society: Peru in Comparative Perspective* (Princeton: Princeton University Press, 1978), pp. 258, 271-73, 280.

[49] For discussion of structure and process in the context of international political economy see C. Fred Bergsten, Robert O. Keohane, and Joseph S. Nye, Jr., "International Economics and International Politics: A Framework for Analysis," in *World Politics and International Economics*, C. Fred Bergsten and Lawrence, eds. (Washington, D.C.: Brookings Institution, 1975); Cheryl Christensen, "Structural Power and National Security," in *Economic Issues and National Security*; and Keohane and Nye, "World Politics and the International Economic System." See also Waltz, Theory of International Politics, pp. 73-101.

[50] Reference is made here to a "second face" of power as introduced in Peter Bachrach and Morton Baratz, "Decisions and Nondecisions: An Analytical Framework," *American Political Science Review*, v. 57 (1963).

[51] A similar distinction has been made in dependency theory between the "external" and "internal" aspects of dependency. See Cardoso and Faletto, *Dependency and Development in Latin America*, pp. xvi-xvii, 22, 27; and Szentes, *The Political Economy of Underdevelopment*, pp. 53-54, 163, 233.

[52] In 1975, exports of non-oil LDCs covered only 67% of their imports. John Holsen and Jean Waelbroeck, *LDC Balance of Payments Policy and the International Monetary System*, World Bank Staff Working Paper, n. 226 (Washington: World Bank, 1976), p. 10.

[53] Moran, *Multinational Corporations and the Politics of Dependence*, p. 97.

[54] Guillermo O'Donnell, *Modernization and Bureaucratic-Authorization: Studies in South American Politics* (Berkeley: Institute of International Studies, 1973); "Reflections on the Patterns of Change in the Bureaucratic-Authoritarian State," *Latin American Research Review*, v. 13, n. 1 (1978).

[55] Stepan, *The State and Society*, pp. 285-87.

56 Evans, *Dependent Development*, p. 271.

57 *Economist*, November 29, 1980.

58 *Economist*, August 1, 1981.

59 Chapters III and IV establish the criteria for the selection of the American machine tool industry for sector case study.

60 On the distinctiveness of issue areas and the difficulties of issue linkages across issue areas, see Keohane and Nye, *Power and Interdependence*.

61 OECD, *Statistics of Foreign Trade, Annual: Tables by Reporting Countries (Series B)*. Manufactured trade: SITC 5-8. 1979 figures.

62 Baldwin, "The Political Economy of Postwar U.S. Trade Policy;" Richard E. Caves, "Economic Models of Political Choice: Canada's Tariff Structure," *Canadian Journal of Economics*, v. 9 (1976); Edward John Ray, "The Optimum Commodity Tariff and Tariff Rates in Developed and Less Developed Countries," *Review of Economics and Statistics*, v. 56 (August 1974); "The Determinants of Tariff and Non-Tariff Trade Restrictions in the United States."

63 The obvious exceptions being producer dependence on capital and raw material imports.

64 The machine tool case study is also chosen for its difficulty and theoretical significance. See Chapters III and IV.

65 Peter J. Katzenstein, "International Relations and Domestic Structures: Foreign Economic Policies of Advanced Industrial States," *International Organization*, v. 30, n. 1 (Winter 1976); "Conclusion: Domestic Structures and Strategies of Foreign Economic Policy," in *Between Power and Plenty*; Stephen D. Krasner, "U.S. Commercial and Monetary Policy: Unraveling the Paradox of External Strength and Internal Weakness," in *Between Power and Plenty*; *Defending the National Interest*.

66 Selections in *Between Power and Plenty*; Peter J. Katzenstein, "Capitalism in One Country?: Switzerland in the International Economy, *International Organization*, v. 34, n. 4 (Autumn 1980); Krasner, *Defending the National Interest*.

[67] Michel Crozier, Samuel Huntington, and Joji Watanuki, *The Crisis of Democracy: Report on the Governability of Democracies to the Trilateral Commission* (New York: New York University Press, 1975), p. 105.

Chapter II

Export Dependence and U.S. Trade Relations

In Chapter I we established that the export dependence thesis could be most fruitfully (although not necessarily most easily) tested with regard to manufactured exports and the United States. Accordingly, in this chapter the effect of export dependence on U.S. commercial policy is examined in an aggregate data analysis of 129 manufacturing sectors. The 129 sectors are based on the three-digit industry groups of the Standard Industrial Classification (SIC) and comprehensively cover the full range of U.S. manufacturing activity as classified by the SIC.[1]

An analytical framework for the aggregate data analysis is established in the following section. The export dependence thesis is first tested against the record of U.S. protectionism as represented by escape clause, antidumping, and countervailing duty cases. A similar test is then conducted regarding voluntary export restraint agreements, orderly marketing arrangements, quotas, the steel trigger price mechanism, and the Multifiber Arrangement. Before concluding, we examine the possibility that other factors and explanations may be responsible for our results.

FRAMEWORK FOR ANALYSIS: THE EXPORT DEPENDENCE - IMPORT PENETRATION MATRIX

The extent to which an industry or entire economy becomes reliant upon exports or any other form of international transaction depends upon the importance of the external link relative to corresponding domestic transactions. Operational indicators of export dependence must then measure the relative importance of foreign versus domestic markets.[2]

Accordingly for each of the 129 industrial sectors, export dependence (*XD*) is operationalized as:

$$XD = \frac{X}{Sp + M}$$

where *X* stands for exports, *Sp* for product shipments, and *M* for imports.[3] Product shipments are total sales in a given product line.

Two major objections have been raised against export-to-production ratios.[4] First, export figures include products utilizing imported inputs. The inclusion of foreign value-added in a country's export data inflates the numerator in export-to-production ratios and exaggerates the importance of trade. This problem is alleviated in the ratio presented above by using a gross output (product shipments) instead of a net output (value-added) base in the denominator. The effect of imported inputs in the shipments data compensates for its effect in the trade data. We can make the assumption, "quite plausible in the case of industrial countries,"[5] that the import content of production for exports is similar to the import content of total production.

A second criticism concerns the fact that some portion of a country's exports are eventually repurchased, in altered form, for domestic consumption. Export figures are again inflated by this by-product of trade in intermediate goods. To discount this second effect, the ratio used here includes imports, and therefore any repurchased exports, in the denominator.

To examine the impact of export dependence on the protectionist behavior of the 129 sectors, sectors are classified by level of export dependence and import penetration in the 2 x 2 matrix displayed in Figure 2.1 (specific procedures are outlined in the following section).

Figure 2.1

Export Dependence-Import Penetration Matrix

	LOW EXPORT DEPENDENCE	HIGH EXPORT DEPENDENCE
LOW IMPORT PENETRATION	I	II
HIGH IMPORT PENETRATION	III	IV

As discussed in the first chapter, import penetration is centrally relevant to the many theses competing with the export dependence argument. To approximate domestic market share captured by foreign suppliers, import penetration (*MP*) is operationalized as:[6]

$$MP = \frac{M}{(Sp + M - X) + X}$$

where *M* stands for imports, *Sp* for product shipments, and *X* for exports. Domestic demand (shipments plus imports minus exports) is used as the base of the ratio in lieu of product-specific measures of domestic consumption. Allowances are made to compensate for two major problems of import-to-production ratios; the obverse of those for export-to-production ratios. Product shipments (gross output) are incorporated into the denominator to compensate for imports which utilize domestic exports (domestic value added). Exports have been included in the denominator to discount those imports that go into production of exports.

The usefulness of the export dependence - import penetration matrix (hereinafter referred to as "*XD-MP* matrix") lies in the anticipation of different protectionist findings across its four cells. For those sectors low on both export dependence and import penetration (Cell I), no specific findings are expected. The export dependence thesis can neither be refuted nor confirmed by the results in this first cell where constraints on protectionism are low. Low levels of protectionism are expected for sectors manifesting high export dependence and low import penetration (Cell II). In this second cell, constraints are high and motivations for protection low. In direct contrast, sectors low on export dependence and highly import penetrated (Cell III) would have the greatest potential for protectionism. Strong protectionist motivations here are reinforced by minimal constraints. The fourth cell of high dependence and high penetration contains "crucial case" sectors for the thesis since the two variables predict contradictory outcomes for protection. To have significant explanatory power, the export dependence thesis would lead one to expect that those highly export dependent sectors in the crucial fourth cell would manifest relatively low levels of protectionism despite being threatened by imports.

In the following two sections, the above hypotheses are examined against the protectionist behavior of U.S. manufacturing sectors as classified by their levels of export dependence and import penetration in the late 1970s. Table 2.1 presents the provisional sector composition of the *XD-MP* matrix for 1975 through 1979. For each sector, levels of export dependence and import penetration were determined for 1977 (the mid-year of the 1975-1979 period) and were raised or lowered by their derived per annum rates of change over 1972 levels. For each of the two variables, the sectors were rank ordered and the weighted mean determined (where one-half of the economy — in terms of shipments — lies above and one-half below the mean). For the 1975-1979 period, the weighted mean for export dependence was 4.76% and for import penetration, 5.25%. The sector composition of the *XD-MP* matrix was determined by classifying those sectors falling below the means as "low" and those at or above the means as "high."

Table 2.1

Provisional 1975-1979 XD-MP Matrix: Sector Composition by Cell, Levels of Export Dependence and Import Penetration by Sector and Cell, and Protectionist Actions* by Sector and Cell (U.S.)

(a) Cell I: low export dependence-low import penetration

SIC-Based Code & Title	XD %	MP %	Protectionist Actions 1975-1979 EC	AD	CVD
201 Meat products	3.90	3.18	4	0	1
202 Dairy products	1.00	1.38	0	0	9
203 Preserved fruits & vegetables	3.56	3.39	5	0	1
205 Bakery products	0.27	0.54	0	0	1
209 Misc foods	2.98	4.95	4	0	0
213 Chewing & smoking tobacco	2.69	4.57	3	0	0
225 Knitting mills	1.00	0.51	3	0	0
227 Floor covering mills	2.78	3.47	0	0	0
228 Yarn & thread mills	2.44	1.44	0	1	3
239 Misc fabricated textile prods	4.32	3.16	2	0	0
244 Wood containers	0.73	1.12	0	0	0
249 Misc wood products	3.75	4.74	4	0	0
25 Furniture & fixtures	1.57	4.04	0	0	0
264 Misc converted paper	2.67	1.17	0	2	0
products	0.61	0.10	0	0	0
265 Paperboard containers	0.07	0.30	0	0	0
271 Newspapers					
272 Periodicals	3.39	0.46	0	0	0
273 Books	4.57	2.29	0	0	0
275 Commercial printing	0.97	0.64	0	0	0
277 Greeting card publishing	0.95	0.57	0	0	0

(continued)

Table 2.1 (continued)

(a) Cell I: low export dependence-low import penetration (continued)

SIC-Based Code & Title		XD %	MP %	Protectionist Actions 1975-1979 EC	AD	CVD
278	Blank books	0.51	2.22	0	0	0
279	Printing trade services	0.26	0.02	0	1	0
284	Soaps, cleaners	2.74	0.76	0	0	0
285	Paints & allied products	2.66	0.13	0	0	0
295	Roofing materials	1.51	1.34	0	0	0
303	Reclaimed rubber	1.30	2.26	0	0	0
306	Fabricated rubber prods nec	3.37	0.89	3	0	0
307	Misc plastics products	3.46	2.30	3	3	0
322	Glass & glassware	3.49	3.21	0	0	0
324	Cement, hydraulic	1.01	3.07	0	2	0
327	Concrete, gypsum	0.38	0.28	0	0	0
332	Iron & steel foundries	4.41	0.29	3	0	1
341	Metal cans	0.75	0.63	0	1	0
344	Fabricated structural metal	4.65	0.72	0	0	0
372	Ship & boat building	3.42	1.13	0	0	0
Cell I Aggregate		2.56	1.81	34	10	16

(continued)

Table 2.1 (continued)

(b) Cell II: high export dependence-low import penetration

SIC-Based Code & Title		XD %	MP %	Protectionist Actions 1975-1979 EC	AD	CVD
204	Grain mill products	7.03	0.41	0	0	0
207	Fats & oils	17.08	4.10	0	0	3
211	Cigarettes	11.32	0.13	0	0	0
222	Weav mills, manmade fiber	4.97	4.23	0	1	0
224	Narrow fabric mills	6.80	2.99	0	0	0
241	Logging, camps	17.62	0.89	0	0	0
263	Paperboard mills	8.93	0.35	0	0	0
266	Building paper	5.95	1.29	0	0	0
282	Plastics materials	9.29	1.73	0	7	1
283	Drugs	11.86	4.75	0	4	3
286	Industrial inorganic chem	10.38	4.44	0	9	0
289	Misc chemical products	16.54	4.15	0	0	0
299	Misc petroleum & coal prods	47.04	3.86	0	0	0
304	Rubber & plastic hose	5.43	4.15	0	0	0
321	Flat glass	9.07	4.36	0	1	5
329	Misc nonmetallic min prods	6.06	3.57	0	0	0
335	Nonferrous rolling	4.76	2.46	0	0	0
343	Plumbing & heating	4.98	2.02	0	0	0
346	Metal forgings	5.49	0.63	0	0	0
348	Ordnance	23.40	1.70	0	0	0
353	Construction machinery	28.34	4.21	0	0	0
358	Refrigeration machinery	15.86	1.44	0	0	0
361	Electric distributing equip	5.22	2.98	0	0	0
364	Electric lighting & wiring	7.52	4.13	0	1	0
372	Aircraft & parts	29.03	3.27	0	0	0
374	Railroad equipment	6.83	1.46	0	0	0
379	Misc transportation equip	13.45	0.90	0	1	0
382	Measuring devices	22.00	4.52	0	0	1
384	Medical instruments	13.81	3.57	0	0	0
395	Pens, pencils	12.34	3.56	0	0	0
Cell II Aggregate		13.28	2.80	0	24	13

Table 2.1 (continued)

(c) Cell III: low export dependence-high import penetration

SIC-Based Code & Title		XD %	MP %	Protectionist Actions 1975-1979 EC	AD	CVD
231	Men's suits & coats	0.14	11.18	0	0	0
232	Men's furnishings	1.81	11.08	0	0	4
233	Women's outer wear	0.79	7.01	0	0	0
234	Women's undergarments	3.30	5.53	0	0	1
235	Hats, caps	4.14	13.03	3	0	0
236	Misc outerwear	1.93	18.68	0	0	2
238	Misc apparel	1.60	32.41	3	0	10
243	Millwork, plywood	2.00	7.38	2	1	0
262	Papermills	2.85	14.16	0	3	0
291	Petroleum refining	0.97	9.05	0	0	0
301	Tires	3.93	10.37	2	1	2
302	Rubber & plastic footwear	0.76	68.97	0	0	3
313	Boot & shoe cut stock	3.39	14.30	0	1	0
314	Footwear, except rubber	0.89	25.52	8	1	7
315	Leather gloves	4.68	24.84	3	0	1
317	Handbags	0.83	29.43	0	0	5
325	Structural clay products	0.85	9.33	0	0	1
328	Cut stone	0.56	7.84	0	0	0
331	Basic steel products	2.14	11.59	20	6	4
333	Primary nonferrous metals	2.15	21.16	0	1	0
339	Misc primary metal products	3.73	5.36	0	0	0
345	Screw machine products	3.37	8.61	5	0	3
375	Motorcycles, bicycles	3.07	48.68	0	0	1
396	Costume jewelry	3.99	18.47	4	0	0
Cell III Aggregate		1.64	11.49	54	17	46

(continued)

Table 2.1 (continued)

(d) Cell IV: high export dependence-high import penetration

SIC-Based Code & Title		XD %	MP %	Protectionist Actions 1975-1979 EC	AD	CVD
221	Weaving mills, cotton	11.24	7.23	0	0	0
229	Misc textile goods	5.54	8.31	0	1	5
237	Fur goods	8.03	11.56	0	0	1
242	Sawmills	5.69	17.19	0	0	0
261	Pulp mills	25.48	33.73	0	0	0
281	Industrial inorganic chem	13.21	11.51	3	4	1
287	Agricultural chemicals	13.55	5.33	0	0	1
311	Leather tanning	10.84	9.89	0	0	0
316	Luggage	6.55	20.63	0	0	0
319	Leather goods nec	6.09	13.29	0	0	0
323	Purchased glass products	9.24	5.34	0	0	1
326	Pottery & related products	4.93	28.69	4	0	0
342	Cutlery, hand tools	6.70	6.00	0	1	1
349	Misc fabricated metal prod	5.17	5.33	7	3	3
351	Turbines	19.83	6.58	0	0	0
352	Agricultural machinery	14.19	8.70	0	0	0
383	Optical instruments	10.21	19.02	0	0	0
385	Ophthalmic goods	7.02	21.57	0	0	0
386	Photographic equipment	14.80	10.35	0	2	0
387	Watches, clocks	11.63	38.34	0	0	0
391	Jewelry, silverware	10.42	33.41	7	0	0
393	Musical instruments	14.05	14.08	0	0	1
394	Toys & sporting goods	5.46	18.27	4	5	0
399	Misc manufactures	6.70	6.28	0	0	0
Cell IV Aggregate		10.88	11.42	31	24	17
Total Manufacturing (N=129)						
Weighted Mean#		4.76	5.25			
Unweighted Mean		7.16	8.68			
Aggregate national average		7.16	7.17			

(continued)

Table 2.1 (continued)

*	affirmative and negative escape clause, antidumping, and CVD determinations and escape clause extensions
#	used for sector classification
XD	export dependence: ratio of exports to shipments plus imports; 1977 adjusted by derived annual average rate of change over 1972
MP	import penetration: ratio of imports to shipments plus imports; 1977 adjusted by derived annual average rate of change over 1972
EC	escape clause
AD	antidumping
CVD	countervailing duty

Sources: Export Dependence and Import Penetration -

U.S., Bureau of the Census, *Annual Survey of Manufactures: Value of Product Shipments*

_____, *U.S. Exports: Domestic Merchandise, SIC-Based Products by World Areas*, FT 610.

_____, *U.S. Exports: SIC Division by SIC-Based Two-Digit, Three-Digit and Four Digit Product Code*, EA 675 (unpublished)

_____, *U.S. Imports: Consumption & General, SIC-Based Products by World Areas*, FT 210

_____, *U.S. Imports: SIC Division by SIC-Based Two-Digit, Three-Digit and Four-Digit Product Code*, IA 275 (unpublished)

Sources: Protectionist Actions -

U.S., Department of the Treasury, *Customs Bulletin*

_____, *Customs Bulletin & Decisions*

U.S., *Federal Register*

U.S., International Trade Commission, *Annual Report*.

_____, *Operation of the Trade Agreements Program*

_____, *USITC Publications*

U.S., Office of the United States Trade Representative, *Trade Action Monitoring System* (unpublished)

U.S., President, *Annual Report of the President of the United States on the Trade Agreement Program*

The impact of the varying levels of export dependence across the four cells and 129 sectors can be examined in the context of major manifestations of the "new protectionism."[7] In particular, the export dependence thesis is tested against the record of some of the most important contemporary import restrictions:[8] escape clause, antidumping and countervailing duty determinations, voluntary export restraints, orderly marketing arrangements, quotas, the steel trigger price mechanism, and the Multifiber Arrangement.

As significant and diverse as the above list may be, the restrictions cited certainly do not exhaust the innumerable guises of the new protectionism. Still, the list would be considered to include among the major forms of the new protectionism.[9] These restraints have an added advantage of being sector- and country-specific in application to match disaggregated levels of export dependence and import penetration.

ESCAPE CLAUSE, ANTIDUMPING, AND COUNTERVAILING DUTY CASES

In this section, the export dependence thesis is tested in five ways: (1) a cross-sectional analysis of 1975-1979 escape clause, antidumping and countervailing duty petitions (i.e., industry protectionsim); (2) disaggregated analyses of the three types of cases; (3) a cross-sectional analysis of the cases that resulted protective action (i.e., government protection); (4) a longitudinal analysis of cases from 1960 to 1979; and (5) an updated examination of 1980 escape clause cases.

As a first test, the protectionist behavior of the four cells in Table 2.1 can be examined in a cross-sectional analysis of escape clause, antidumping and countervailing duty determinations from 1975 to 1979. Escape clause (or safeguard) cases involve injury to an entire industry resulting from increased imports.[10] Cases are usually petitioned by trade associations although some have been initiated by Congress and the executive branch. Investigations are conducted by the International Trade Commission (ITC). If no injury is found or if imports are not determined to be "a substantial cause" of injury, the case is terminated. If both the injury and substantial cause determinations are made in the affirmative, the ITC recommends a remedy and the case is sent to the President. The President may accept, modify or reject the ITC recommendation. In cases of modification or rejection, the President's decision is subject to congressional override.

Antidumping cases involve injurious imports that are sold at less than fair value (SLFV).[11] These cases are petitioned by trade associations, groups of firms and individual firms.[12] Since 1979, the International Trade Administration (ITA) of the Commerce Department determines whether sales are being made at less than fair value. (Prior to 1979, SLFV determinations were made by the Customs Service of the Treasury Department.) If an SLFV determination is affirmative, the case is sent to the ITC for a material injury determination. Antidumping duties are imposed if both the ITA determines the existence of dumping and the ITC finds material injury to industry.

Countervailing duty (CVD) cases involve injurious imports that are subsidized by foreign governments.[13] Cases are initiated by trade associations, groups of firms, and, occasionally, individual firms. Investigation of the existence of foreign subsidy is made by Commerce's ITA. (Again, prior to 1979 subsidy determinations were made by Treasury's Customs Service.) Simultaneously, material injury investigations are conducted by the ITC for cases involving duty-free goods, signatories of the Tokyo Round's subsidy/CVD code, and countries to whom the U.S. has extended most favored nation status (most of U.S. imports fall under these conditions). CVD duties are levied if subsidy and injury determinations are both made in the affirmative, or in those few cases where affirmative subsidy findings are made and injury tests are not required.

There has been some doubt as to whether antidumping and CVD actions should be considered protectionist. They are, in principle, legitimate counter-measures taken against unfair trade practices. However, GATT officials Blackhurst, Marian and Tumlir have recognized that

> in a period of generally rising protectionist pressure, antidumping and countervailing procedures may also be abused for protectionist purposes.[14]

Japan has protested U.S. antidumping duties on their consumer electronics as being outright protectionist. The U.S. has made similar charges regarding European CVDs levied against U.S. synthetic fibers. On the basis of both the reality and perception of their intent, antidumping and CVD actions are included here as protectionist measures.[15] Escape clause measures are clearly protectionist since they involve restrictions against injurious imports that result from *fair* trade.

Affirmative escape clause, antidumping and CVD determinations raising tariff or non-tariff restrictions would be appropriate indicators of "protection" — instances where industries are protected from imports. However, "protectionism" or the protectionist tendencies of industry would be better measured by the total number of (completed) cases brought before the government by industry and not just by those cases where protective measures were taken. Since we are interested in explaining not just official protection but underlying protectionist pressures as well, negative as well as affirmative determinations are considered here.

Figure 2.2 presents the provisional protectionist outcomes for the 1975-1979 *XD-MP* matrix (corresponding directly with the provisional sector composition in Table 2.1) in terms of all escape clause, antidumping and CVD determinations (affirmative, negative and escape clause extensions) from 1975 to 1979. The levels of protectionism in each cell were arrived at in a four step process. First, the number of protectionist actions for each industrial sector and then cell were determined (see Table 2.1). Antidumping and CVD petitions explicitly name an allegedly offending country, and each case is considered a single protectionist action. Escape clause motions are industry-wide matters that concern competing imports from all sources. To avoid equating antidumping or CVD actions which affect a single country per case with an escape clause action that may concern many foreign suppliers, each major supplier that may be affected by an escape clause case is considered as being subject to a separate protectionist action.[16]

Second, each action is weighted by the size (in terms of shipments) of the industry concerned. This weighting avoids the problem of equating a protectionist action by the automobile industry with one in the greeting card industry. Protectionist actions are weighted by shipments and not by imports for conceptual and methodological reasons. Conceptually, throughout this chapter we are less interested in the international impact of protectionist actions than with the importance or significance of the industries undertaking protectionist actions. While weighting by imports could measure the foreign effect of protection, weighting by industry size measures the extent of protectionism within an economy.

Methodologically, weighting by imports would seriously bias the results against the two minimally import penetrated cells (I and II). The level of imports across the four cells varies widely and this variation would have a substantial impact on the protectionist results if weighted by imports.

Figure 2.2

Provisional 1975-1979 XD-MP Matrix: Protectionism* by Cell (U.S.)

	LOW EXPORT DEPENDENCE	HIGH EXPORT DEPENDENCE
LOW IMPORT PENETRATION	I 95.4	II 72.8
HIGH IMPORT PENETRATION	III 247.9	IV 72

* weighted affirmative and negative escape clause, antidumping, and countervailing duty determinations, and escape clause extensions; adjusted to the number of cases in cell IV

Source: see Table 2.1

When in the fifth chapter the matrix analysis is dropped and attention turns to the targets of protection, protectionist actions are weighted by imports.

Continuing on with the third step in aggregating cell results, the weighted levels of protectionism were adjusted to compare directly with the total number of actions in the crucial fourth cell to provide better perspective. The number of actions in Cell IV was taken as its weighted level of protectionism and the weighted levels of the other three cells were adjusted accordingly.

Finally, each cell's protectionism is weighted by the proportion of the economy (in terms of shipments) represented in each cell — again taking Cell IV as the base around which all other cells were adjusted. Without this adjustment, a cell's level of protection is influenced by its economic size. This final weighting permits the direct comparison of protectionist outcomes across the four cells.

While the results in Figure 2.1 are to be subsequently adjusted by the following sector-by-sector analysis, a few comments on the provisional cell outcomes can be made. The totals are very supportive of the export dependence argument. As expected, among the lowest levels of protectionism is found in Cell II (72.8), and the highest is found in Cell III (247.9). Protectionism in Cell IV (72), the crucial case cell, is as low as Cell II's. This finding is significant since Cell IV faces as great a threat from import penetration as Cell III (11.4% and 11.5%, respectively). Cell IV's protectionism even falls below that of Cell I (95.4) — a cell not threatened by imports. More will be said of the (adjusted) cell outcomes after an analysis of sector-specific results. (The disaggregated results can be found in Table 2.1.) The purpose of this analysis is not just to specify the findings by sector, but to refine and disaggregate the data yet further to arrive at more valid results.

Cell I. The first cell of low dependence and low penetration manifests the second highest level of protectionism. Cell I sectors with significant protectionist activity (at least one escape clause motion and/or three antidumping/CVD actions) include:

- meat products (SIC 201)
- dairy products (202)
- preserved fruits and vegetables (203)
- miscellaneous foods (209)
- tobacco (213)
- knitting mills (225)
- yarn and thread mills (228)
- miscellaneous fabricated textile products (239)
- miscellaneous wood products (249)
- fabricated rubber products, nec (306)
- miscellaneous plastics products (307)
- iron and steel foundries (332)

Major protective measures were gained by the miscellaneous wood products industry (3.8% *XD*; 4.7% *MP*) when escape clause actions were taken against imported clothespins in 1979. Major suppliers affected were China, Taiwan, Poland and West Germany.

The export dependence thesis cannot explain the protectionism found in this first cell. However, its low level of export dependence (2.6%) presented little obstacle to the protectionism found here whatever its source.

Cell II. Of the thirty sectors initially falling into this second cell of high dependence and low penetration, not a single industry group undertook an escape clause motion between 1975 and 1979 — a period which witnessed 119 such actions. Twenty-one of these sectors did not even initiate an antidumping or CVD investigation. Four other sectors exhibited only minimal protectionist activity (no escape clause motions, not more than two antidumping/CVD actions). The five remaining sectors account for thirty-three of the thirty-seven actions in this cell:

- fats and oils (SIC 207)
- plastics materials and synthetics (282)
- drugs (283)
- industrial organic chemicals (286)
- flat glass (321)

These results are already generally supportive of the expectations for Cell II. Yet further specification and disaggregation of the five cited sectors further reveals the explanatory power of the export dependence variable. Disaggregating export dependence and import penetration at the SIC four- and five-digit levels could disclose that certain subsectors have trade characteristics warranting classification into cells other than that of the full three-digit sector.[17]

Table 2.2 displays the disaggregated export dependence and import penetration levels for the five sectors cited above. Disaggregation of the fats and oils industry group reveals that the shortening and cooking oils industry (SIC 2079) has Cell III not Cell II characteristics: 3.5% *XD*, 11.4% *MP*. Significantly, this subsector accounted for all three protectionist actions of the parent sector. Regarding the plastics materials and synthetics sector, the synthetic rubber (2822) and cellulosic manmade fiber (2823) industries are also reclassified out of Cell II. The reclassification removes six of the parent sector's eight antidumping/CVD actions. Similarly, all seven of the drug sector's protectionist actions are eliminated from Cell II with the reclassification of medicinals and botanicals (2833) and pharmaceuticals (2834).

Still, three subsectors exhibit two or more antidumping/CVD actions:

- plastics and resins (SIC 2821)
- industrial organic chemicals (2869)
- flat and sheet glass (32115)

Table 2.2

SIC Four- & Five-Digit Disaggregation for Selected Cell II Sectors: Levels of Export Dependence & Import Penetration, Protectionist Actions*, and Cell Classification# (U.S.)

SIC-Based Code & Title		XD	MP	Protectionist Actions 1975-1979			CELL
		%	%	EC	AD	CVD	
207	Fats & oils						
2074	Cottonseed oil mills	30.76	0.60	0	0	0	II
2075	Soybean oil mills	22.20	0.09	0	0	0	II
20761	Linseed oil	6.43	0.01	0	0	0	II
2077	Animal & marine fats	29.61	1.43	0	0	0	II
2079	Shortening & cooking oils	3.45	11.37	0	0	3	III
282	Plastics materials						
2821	Plastics materials & resins	8.50	0.74	0	2	0	II
2822	Synthetic rubber	14.22	5.76	0	1	0	IV
2823	Cellulosic manmade fiber	7.56	6.18	0	4	1	IV
2824	Noncellulosic manmade fiber	5.66	1.28	0	0	0	II
283	Drugs						
2831	Biological products	13.15	0.56	0	0	0	II
2833	Medicinals & botanicals	37.77	19.97	0	4	0	IV
2834	Pharmaceutical preparations	3.39	0.38	0	0	3	I

(continued)

Table 2.2 (continued)

SIC-Based Code & Title		XD	MP	Protectionist Actions 1975-1979			CELL
		%	%	EC	AD	CVD	
286	Industrial organic chemicals						
2861	Gum & wood chemicals	13.88	4.11	0	0	0	II
2865	Cyclic crudes	15.24	8.74	0	0	0	IV
2834	Industrial organic chem nec	8.77	3.10	0	9	0	II
321	Flat glass						
32113	Laminated glass	6.16	5.68	0	0	0	IV
32115	Flat & sheet glass	12.11	3.64	0	1	5	II

* affirmative & negative escape clause, antidumping, & countervailing duty determinations, and escape clause extensions

\# based on weighted means of 4.76% XD & 5.25% MP for three-digit industry groups

Sources: see Table 2.1

Yet further disaggregation of these industries and application of the export dependence thesis at more specific levels can explain their protectionism. In this case, an industry's trade position can be broken down to determine dependence on and penetration by two groups of countries: industrial countries and the rest-of-the-world. A given industry may be highly dependent on one of these markets but not the other — translating into different degrees of constraint on protectionism against the two groups of countries.

Disaggregating the 129 SIC-based three-digit industry groups by their 1977 trade with industrial countries versus the rest of the world reveals 2.33% and 1.65% as the weighted means for export dependence and import penetration, respectively.[18] Using these levels as criteria for cell classification, we can now examine the disaggregated results for the three subsectors cited above (see Table 2.3).

Table 2.3 reveals that the industrial organic chemicals industry's trade relations with the industrial world (5.1% *XD*, 2.7% *MP*) is to be reclassified out of Cell II. Seven of the industry's nine protectionist actions were aimed at industrial countries. Similarly, the flat and sheet glass industry's trade relations with industrial countries do not have Cell II characteristics (7.3% *XD*, 3.3% *MP*). And this segment of the industry was responsible for five of the industry's six protectionist actions.

Disaggregation at the SIC four- and five-digit levels and between trade relations with industrial countries versus the rest-of-the-world strengthens what were already good results for Cell II. Overall, twenty-eight of the thirty-seven protectionist actions initially found in this cell are reclassified to other cells.

Table 2.3

Disaggregation of Selected Cell II Subsectors by Trade with Industrialized Countries & the Rest-of-the-World: Levels of Export Dependence & Import Penetration*, Protectionist Actions#, and Cell Classification@ (U.S.)

SIC-Based Code & Title		XD %	MP %	Protectionist Actions 1975-1979 EC	AD	CVD	CELL
2821	Plastics materials & resins						
	trade w/ industrial world	4.71	0.76	0	2	0	II
	trade w/ ROW	3.87	0.02	0	0	0	II
2869	Industrial organic chem nec						
	trade w/ industrial world	8.50	0.74	0	7	0	IV
	trade w/ ROW	14.22	5.76	0	2	0	II
32115	Flat & sheet glass						
	trade w/ industrial world	7.26	3.28	0	0	5	IV
	trade w/ ROW	2.79	1.13	0	1	0	II

* 1977 levels only

\# affirmative & negative escape clause, antidumping, & countervailing duty determinations, and escape clause extensions

@ based on weighted means of 2.33% XD & 1.65% MP when disaggregating trade relations at the SIC three-digit level

ROW rest-of-the-world

Sources: see Table 2.1

Cell III. The third cell of low dependence and high penetration was expected to contain the most protectionist sectors of the economy and we are not disappointed. Thirteen Cell III sectors exhibited significant protectionist activity:

- sugar and confectionary products (SIC 206)
- men's and boy's furnishings (232)
- hats (235)
- miscellaneous apparel (238)
- plywood millwork (243)
- paper mills (262)
- rubber and plastic footwear (302)
- nonrubber footwear (314)
- leather gloves and mittens (315)
- handbags (317)
- basic steel products (331)
- screw machine products (345)
- costume jewelry (396)

Among the most noteworthy protective measures of the late 1970s are encompassed in Cell III. In the basic steel industry (2.1% *XD*, 11.6% *MP*), escape clause protection was invoked against imports of stainless steel and alloy tool steel in 1976. Japan, the European Community, and Sweden were the major targets of these measures. After one escape clause motion against imported high-carbon ferrochromium was denied by the President in January 1978, safeguard duties were imposed in a second case that November. South Africa, Rhodesia, Brazil and Japan were most affected by the five-year increased tariff, Carbon steel plate from Japan and Taiwan became subject to antidumping duties in 1978 and 1979, respectively. Other protectionist petitions that have been filed by the steel industry but denied include: round stainless steel wire (escape clause 1976), low-carbon ferrochromium (escape clause 1977), welded stainless steel pipe and tube from Japan (antidumping 1978), coat hangars from Canada, carbon steel plate from Poland, coke from West Germany (antidumping 1979), carbon steel plate from Mexico, ferrochromium from South Africa (CVD 1976), and silicon electrical steel from Italy (CVD 1977, 1979).

In 1975 and 1977 escape clause relief for the screw machine products sector (3.4% *XD*, 8.6% *MP*) was rejected. In yet a third petition in December 1978, the industry gained a three year increase in tariffs against imported bolts, nuts and screws (major suppliers: Japan,

Canada). In addition, the industry secured CVD duties against Italy (1976) and Japan (1977, 1979).

Safeguard protection for the nonrubber footwear industry (0.9% *XD*, 25.5% *MP*) was instituted in 1977. Between 1976 and 1979, four CVDs were imposed against shoe imports from South Korea, Spain, Argentina and India. CVD motions against Argentina (a second case) and Uruguay were denied as was an antidumping petition against Rumania.

In 1978, the sugar industry (1.4% *XD*, 14.1% *MP*) successfully petitioned for CVD duties against the European Community; targeting mainly French and German exports. The following year, in three separate actions, antidumping duties were levied against sugar from Belgium, France and West Germany. An escape clause motion by the industry had failed in 1977.

As was the case with Cell II, expectations were more than fulfilled for Cell III.

Cell IV. Despite being highly penetrated by imports, twenty-eight of the thirty-six sectors provisionally classified in the fourth cell exhibited minimal or absolutely no protectionist activity. Of these sectors, which petitioned no escape clause cases and not more than two antidumping/CVD cases, twelve are characterized by import penetration levels in excess of 10%: watches (38.3%), pulp mills (33.7%), ophthalmic goods (21.6%), luggage (20.6%), optical instruments (14.1%), special industrial machinery (14.0%), leather goods, nec (13.3%), fur goods (11.6%), photographic equipment (10.4%).

Eight other industry groups with at least one escape clause action and/or three or more antidumping/CVD motions stand as seeming sore points in this crucial cell:

- miscellaneous textile goods (SIC 229)
- industrial inorganic chemicals (281)
- pottery and related products (326)
- miscellaneous fabricated metal products (349)
- general industrial machinery (356)
- radio and television receiving equipment (365)
- jewelry, silverware and plated ware (391)
- toys and sporting goods (394)

These eight sectors are responsible for 100% of the escape clause and 68% of the antidumping /CVD actions provisionally found in Cell IV. A full 75% of these actions can, however, be accounted for by disaggregating the sectors along the lines in the analysis of Cell II. Table 2.4 presents the trade positions of the eight "trouble" sectors disaggregated at the SIC four- and five-digit levels. In light of notable trade actions concerning the motor vehicle industry, SIC 371 is also disaggregated. The breakdown reveals that a number of protectionist subsectors can be reclassified out of Cell IV.

In the miscellaneous textile good sector, all but the lace goods industry (SIC 2292) is transferred out of the fourth cell. With the transfer go three antidumping actions in the coated fabric (2295), cordage and twine (2298) and textile goods, nec (2299) industries. The three remaining protectionist actions in the three-digit sector cannot be classified at the four- or five-digit level since the cases were not aimed at specific product lines. One of the antidumping cases in the general industrial machinery sector targeted water circulating pumps from Britain. As shown in Table 2.4 this action and the pumps and pumping equipment industry (SIC 3561) are reclassified out of Cell IV.

The most significant results of the disaggregated analysis concern the miscellaneous fabricated metal products, radio and television receiving equipment, and jewelry and silverware industry groups. Regarding the fabricated metal products sector, two subsectors can be removed from Cell IV: miscellaneous fabricated wire products (SIC 3496) with Cell III characteristics (3.6% *XD*, 10.2% *MP*) and fabricated metal products, nec (3499) with Cell I attributes (3.6% *XD*, 3.8% *MP*). The former was responsible for six antidumping/CVD motions, the latter for seven escape clause actions — or the total protectionist activity of the parent sector. Significantly, the one remaining industry exhibiting Cell IV characteristics, valves and pipe fittings (3494), sought no import protection.

While the three-digit radio and television industry group is both highly export dependent and import penetrated (6.2% *XD*, 39.0% *MP*), substantial portions of the sector do not belong in Cell IV. The radio industry (SIC 36511) is minimally dependent on exports (2.5%) while severely penetrated by imports (66.5%). The television industry (36512) has similar Cell II characteristics (2.5% *XD*, 27.6% *MP*). These two Cell III industries are responsible for all six of the parent sector's escape clause action.

Table 2.4

SIC Four- & Five-Digit Disaggregation for Selected Cell IV Sectors: Levels of Export Dependence & Import Penetration, Protectionist Actions*, and Cell Classification# (U.S.)

SIC-Based Code	& Title	XD	MP	Protectionist Actions 1975-1979			CELL
		%	%	EC	AD	CVD	
229	Misc textile goods						
2291	Felt goods	2.60	1.22	0	0	0	I
2292	Lace goods	13.87	27.34	0	0	0	IV
2295	Coated fabrics	9.00	1.53	0	1	0	II
2296	Tire cord & fabric	7.40	0.32	0	0	0	II
2297	Nonwoven fabrics	4.33	2.73	0	0	0	I
2298	Cordage & twine	3.53	21.80	0	0	1	III
2299	Textile goods nec	3.71	35.67	0	0	1	III
281	Industrial inorganic chems						
2812	Alkalies & chlorine	10.16	2.20	0	0	0	II
2813	Industrial gases	16.80	11.04	0	0	0	IV
2816	Inorganic pigments	4.86	12.89	3	4	0	IV
2819	Industrial inorganic chems	12.85	16.31	0	0	1	IV

(continued)

Table 2.4 (continued)

SIC-Based Code & Title		XD %	MP %	Protectionist Actions 1975-1979 EC	AD	CVD	CELL
326	Pottery & related products						
3261	Vitreous plumbing fixtures	5.16	1.72	0	0	0	II
3262	Vitreous china utensils	2.52	43.66	0	0	0	III
3263	Fine earthenware utensils	1.85	62.82	0	0	0	III
3264	Porcelain electrical suppls	7.63	7.50	0	0	0	IV
3269	Pottery products nec	5.71	43.79	4	0	0	IV
349	Misc fabricated metal prods						
3493	Steel springs	2.48	13.15	0	0	0	II
3494	Valve & pipe fittings	13.44	5.65	0	0	0	IV
3496	Misc fabricated wire prods	3.63	10.24	0	3	3	II
3497	Metal foil & leaf	1.69	6.05	0	0	0	III
3499	Fabricated metal prods nec	3.57	3.84	7	0	0	I

(continued)

Table 2.4 (continued)

SIC-Based Code	& Title	XD %	MP %	Protectionist Actions 1975-1979 EC	AD	CVD	CELL
356	General industrial machinery						
3561	Pumps	16.70	4.90	0	1	0	II
3562	Ball & roller bearings	7.62	10.00	0	1	0	IV
3563	Air & gas compressors	24.70	6.16	0	0	0	IV
3564	Blowers & fans	30.49	8.65	0	1	0	IV
3566	Speed changers	4.31	5.56	0	0	0	III
3567	Industrial furnaces & ovens	20.89	3.66	0	0	0	II
3568	Power transmission equip	11.37	3.83	0	0	0	II
365	Radio & television equipment						
36511	Radios	2.47	66.54	3	0	0	III
36512	Television receivers	2.51	27.63	3	0	0	III
36514	Recorders, phonographs	13.49	49.50	0	1	2	IV
36520	Phonograph records	7.12	2.58	0	0	0	II
371	Motor vehicles & equipment						
37111	Passenger cars	2.54[a]	18.53	0	0	0	III
37112	Trucks, buses	4.82[a]	8.32	0	0	0	IV
3714	Motor vehicle parts	6.54[a]	8.72	0	0	0	IV

(continued)

Table 2.4 (continued)

SIC-Based Code & Title		XD %	MP %	Protectionist Actions 1975-1979 EC	AD	CVD	CELL
391	Jewelry, silverware						
3911	Jewelry	11.41	34.75	0	0	0	IV
3914	Silverware & plated ware	1.67	21.68	7	0	0	III
394	Toys & sporting goods						
3942	Dolls	1.79	21.59	0	0	0	III
3944	Games, toys	5.69	15.18	0	0	0	IV
3949	Sporting & athletic goods	6.68	19.11	4	5	0	IV

* affirmative & negative escape clause, antidumping, & countervailing duty determinations, and escape clause extensions

based on weighted means of 4.76% XD & 5.25% MP for three-digit industry groups

a excluding estimated exports under the US-Canadian Automotive Products Trade Agreement

Sources: see Table 2.1

Similarly revealing findings result from the disaggregation of the jewelry, silverware and plated ware industry group. This three-digit sector, which in the aggregate has Cell IV characteristics (10.4% *XD*, 33.4% *MP*), encompasses two very different industries. The jewelry industry (SIC 3911) is, indeed, highly dependent (11.4%) as well as highly penetrated (34.8%). Significantly, this industry exhibits no protectionist activity. In contrast, the silverware and plated ware industry (3914) is scarcely dependent on exports (1.7%) but is seriously import penetrated (21.7%). This Cell III subsector accounts for all seven of the parent sector's (escape clause) actions.

Seven industries still exhibit one or more escape clause actions and/or two or more antidumping/CVD actions:

- miscellaneous textile goods (SIC 229)
- inorganic pigments (2816)
- cellulosic manmade fibers (2823) (transferred from Cell II)
- medicinals and botanical drugs (2833) (transferred from Cell II)
- pottery products, nec (3269)
- phonograph and recorder systems (36514)
- sporting goods (3949)

Much of the protectionist activity evidenced by these industries can, again, be accounted for by distinguishing between their trade relations with industrial countries versus the rest-of-the-world (see Table 2.5).

Three of the six protectionist actions in the miscellaneous textile goods sector have already been reclassified out of Cell IV with the SIC four- and five-digit analysis. The three remaining CVD cases could not be classified at such specific levels, but did target imports from Brazil and India (twice). As shown in Table 2.5, U.S. trade relations with such developing countries in textile goods falls in Cell III (1.8% *XD*, 6.4% *MP*). No actions were aimed at industrial countries with which U.S. trade relations fall in Cell IV (3.4% *XD*, 2.6% *MP*). As a result of the disaggregated analysis, all six of this three-digit sector's protectionist actions are removed from Cell IV.

Table 2.5

Disaggregation of Selected Cell IV Subsectors by Trade with Industrialized Countries & the Rest-of-the-World: Levels of Export Dependence & Import Penetration*, Protectionist Actions#, and Cell Classification@ (U.S.)

SIC-Based Code	Title	XD %	MP %	Protectionist Actions 1975-1979 EC	AD	CVD	CELL
229	Misc textile goods						
	trade w/ industrial world	3.43	2.57	0	0	0	IV
	trade w/ ROW	1.76	6.40	0	0	3	III
2816	Inorganic pigments						
	trade w/ industrial world	2.96	9.06	3	4	0	IV
	trade w/ ROW	1.75	1.93	0	0	0	III
229	Cellulosic manmade fiber						
	trade w/ industrial world	1.73	5.41	0	3	1	III
	trade w/ ROW	5.10	0.25	0	1	0	II
229	Medicinals & botanicals						
	trade w/ industrial world	27.30	17.37	0	3	0	IV
	trade w/ ROW	9.72	2.92	0	1	0	IV

(continued)

Table 2.5 (continued)

SIC-Based Code & Title		XD %	MP %	Protectionist Actions 1975-1979 EC	AD	CVD	CELL
229	Pottery products nec						
	trade w/ industrial world	2.85	32.02	2	0	0	IV
	trade w/ ROW	2.10	0.51	2	0	0	I
229	Recorders, phonographs						
	trade w/ industrial world	6.52	41.23	0	1	2	IV
	trade w/ ROW	6.00	9.74	0	0	0	IV
32115	Sporting & athletic goods						
	trade w/ industrial world	5.28	9.30	1	4	0	IV
	trade w/ ROW	1.91	8.63	3	1	0	III

* 1977 levels only

\# affirmative & negative escape clause, antidumping, & countervailing duty determinations, and escape clause extensions

@ based on weighted means of 2.33% XD & 1.65% MP when disaggregating trade relations at the SIC three-digit level

ROW rest-of-the-world

Sources: see Table 2.1

The Cell IV characteristics of the cellulosic manmade fibers industry (producers of rayon and acetate yarn) mask two very different segments of the industry. The industry is highly penetrated by industrial country suppliers (5.4%) while only minimally dependent on industrial country markets (1.7%). This Cell III segment of the industry manifested three antidumping actions against Belgium, France, and Italy, and a CVD motion against Sweden. In stark contrast, the industry faces virtually no import penetration from outside the industrial world (0.3%) and is highly dependent on these other markets (5.1%) — Cell II characteristics. Consequently, the entire four-digit industry and its five protectionist actions are transferred out of Cell IV.

Outside of the industrial world, the pottery industry is neither export dependent (2.1%) nor import penetrated (0.5%). Hence the two developing country targets (Brazil and South Korea) of the industry's 1976 escape clause motion drop out of Cell IV and are transferred to Cell I.

Finally, while the sporting goods industry's trade relations with industrial countries has Cell IV characteristics (5.3% *XD*, 9.3% *MP*), the industry is minimally dependent (1.9%) on and highly penetrated (8.6%) by the rest-of-the-world. This Cell III segment of the industry sought escape clause protection against Taiwanese, Mexican and South Korean fishing tackle, and antidumping duties against Finnish ice hockey sticks in 1978.

Overall, the analysis at disaggregated levels eliminates forty-four (twenty-five escape clause) of the seventy-two protectionist actions initially found in Cell IV.

Extension of the export dependence thesis at disaggregated levels significantly improves what were already solid results. (The adjusted sector composition and protectionist activity of the 1975-1979 *XD-MP* matrix is presented in Table 2.6,) Five industries and twelve protectionist actions (nine escape clauses) were added to the first cell. Among them, the pharmaceutical, pottery (regarding trade outside the industrial world) and fabricated metal products (nec) industries gained notable trade protection. Between 1976 and 1979, the drug industry secured the imposition of CVD duties on vitamins, ampicillin tryhydrate, and amoxicillin tryhydrate all from Spain. In 1976 the pottery industry gained escape clause protection against earthenware table and kitchen articles. Brazil and South Korea were among the major targets of the three-year increase in tariff rates. In 1979, the fabricated metal products industry successfully petitioned for safeguard protection against imported porcelain-on-steel cookware. The four-year increase in tariffs most affected Japan, Taiwan, Mexico and Italy.

Table 2.6

Final 1975-1979 XD-MP Matrix: Sector Composition by Cell, Levels of Export Dependence and Import Penetration by Cell, and Protectionist Actions* by Sector and Cell 1960-1979 (U.S.)

(a) Cell I: low export dependence (2.56%)-low import penetration (1.88%)

SIC-Based Code & Title		Protectionist Actions			
		1960-1964	1965-1969	1970-1974	1975-1979
201	Meat products	3	0	1	4
202	Dairy products	0	0	0	0
203	Preserved fruits & vegetables	1	1	2	5
205	Bakery products	0	0	1	0
209	Misc foods	0	4	0	4
213	Chewing & smoking tobacco	0	0	0	3
225	Knitting mills	0	0	0	3
227	Floor covering mills	2	6	0	0
228	Yarn & thread mills	0	0	0	1
2291	Felt goods	0	0	0	0
2297	Nonwoven fabrics	0	0	0	0
239	Misc fabricated textile prod	5	0	1	2
244	Wood containers	0	0	0	0
249	Misc wood products	1	0	1	4
25	Furniture & fixtures	0	0	0	0
264	Misc converted paper prods	0	0	0	2
265	Paperboard containers	0	0	0	0
271	Newspapers	0	0	0	0
272	Periodicals	0	0	0	0
273	Books	0	0	0	0
275	Commercial printing	0	0	0	0
277	Greeting card publishing	0	0	0	0
278	Blank books	0	0	0	0
279	Printing trade services	0	0	0	1

(continued)

Table 2.6 (continued)

SIC-Based Code	& Title	Protectionist Actions 1960-1964	1965-1969	1970-1974	1975-1979
2834	Pharmaceutical preparations	0	0	1	0
284	Soaps, cleaners	0	0	0	0
285	Paints & allied products	0	0	0	0
295	Roofing materials	0	0	0	0
303	Reclaimed rubber	0	0	0	0
306	Fabricated rubber prods nec	0	1	0	3
307	Misc plastics products	2	0	3	6
322	Glass & glassware	0	0	0	0
324	Cement, hydraulic	6	0	0	2
3269-R	Pottery, products nec	0	0	0	2
327	Concrete, gypsum	0	0	0	0
332	Iron & steel foundries	2	1	1	3
341	Metal cans	0	0	0	1
344	Fabricated structural metal	0	0	1	0
3499	Fabricated metal prods nec	1	0	2	7
373	Ship & boat building	0	0	0	0
Cell I Aggregate		23	13	14	53

(b) Cell II: high export dependence (13.56%)-low import penetration (2.23%)

SIC-Based Code	& Title	Protectionist Actions 1960-1964	1965-1969	1970-1974	1975-1979
204	Grain mill products	1	0	0	0
2074	Cottonseed oil mills	0	0	0	0
2075	Soybean oil mills	0	0	0	0
20761	Linseed oil	0	0	0	0
2077	Animal & marine fats	0	0	0	0
211	Cigarettes	0	0	0	0
222	Weaving mills, manmade fiber	0	0	1	1

(continued)

Table 2.6 (continued)

SIC-Based Code & Title		Protectionist Actions			
		1960-1964	1965-1969	1970-1974	1975-1979
224	Narrow fabric mills	0	0	0	0
2295	Coated fabrics	0	0	0	1
2296	Tire cord & fabric	0	0	0	0
241	Logging, camps	0	0	0	0
263	Paperboard mills	0	0	0	0
266	Building paper mills	0	0	0	0
2812	Alkalies & chlorine	0	0	0	0
2821	Plastics materials & resins	0	0	1	2
2823-R	Cellulosic manmade fiber	1	0	0	1
2824	Noncellulosic manmade fiber	0	0	0	0
2831	Biological products	0	0	0	0
2861	Gum & wood chemicals	0	0	0	0
2869-R	Industrial organic chem nec	0	0	0	2
289	Misc chemical products	0	0	0	0
299	Misc petroleum & coal prods	0	0	0	0
304	Rubber & plastic hose	0	0	0	0
32115-R	Flat & sheet glass	3	1	1	1
3261	Vitreous plumbing fixtures	0	0	0	0
329	Misc nonmetallic minl prod	0	0	1	0
335	Nonferrous rolling	2	0	0	0
343	Plumbing & heating	0	0	0	0
346	Metal forgings	0	0	1	0
348	Ordnance	0	0	0	0
353	Construction machinery	0	0	3	0
3561	Pumps	0	0	1	1
3567	Industrial furnaces	0	0	0	0
3568	Power transmission equip	0	0	0	0
358	Refrigeration machinery	0	0	1	0
361	Electrical distributing equip	0	0	5	0

(continued)

Table 2.6 (continued)

SIC-Based Code & Title		Protectionist Actions			
		1960-1964	1965-1969	1970-1974	1975-1979
364	Electric lighting & wiring	0	0	0	1
36520	Phonograph records	0	0	0	0
372	Aircraft & parts	0	0	0	0
374	Railroad equipment	0	0	0	0
379	Misc transportation equip	0	0	0	1
382	Measuring devices	1	0	1	0
384	Medical instruments	0	0	0	0
395	Pens, pencils	3	0	0	0
Cell II Aggregate		11	1	16	11

(c) Cell III: low export dependence (1.84%)-high import penetration (13.07%)

SIC-Based Code & Title		Protectionist Actions			
		1960-1964	1965-1969	1970-1974	1975-1979
206	Sugar products	0	0	0	7
2079	Shortening & cooking oils	0	0	0	0
208	Beverages	1	0	0	0
212	Cigars	0	0	0	0
223	Weaving mills, wool	0	0	1	0
2298	Cordage & twine	2	0	1	0
2299	Textile goods nec	0	0	0	0
229-R	Misc textile goods	0	0	0	0
231	Men's suits & coats	0	0	0	0
232	Men's furnishings	0	0	0	0
233	Women's outerwear	0	0	0	0
234	Women's undergarments	0	0	0	0
235	Hats, caps	0	0	0	3
236	Misc outerwear	0	0	0	0
238	Misc apparel	1	0	0	3
243	Millwork, plywood	1	0	0	3

(continued)

Table 2.6 (continued)

SIC-Based Code & Title		Protectionist Actions			
		1960-1964	1965-1969	1970-1974	1975-1979
262	Papermills	0	0	2	3
2816-R	Inorganic pigments	0	0	0	0
2823-I	Cellulosic manmade fibers	4	0	0	3
291	Petroleum refining	0	0	0	0
315	Leather gloves	2	0	0	3
317	Handbags	0	0	0	0
325	Structural clay products	1	0	2	0
3262	Vitreous china utensils	1	0	0	0
3263	Fine earthenware utensils	0	0	0	1
328	Cut stone & stone products	0	0	2	0
331	Basic steel products	6	4	7	26
333	Primary nonferrous metals	0	0	2	1
339	Misc primary metal products	0	0	1	0
345	Screw machine products	0	0	0	5
3493	Steel springs	0	0	0	0
3496	Misc fabricated wire prods	1	0	3	3
3497	Fabricated metal prods nec	0	0	0	0
3566	Speed changers	0	0	0	0
36511	Radios	0	0	0	3
36512	Television receivers	0	0	2	3
37111	Passenger cars	0	0	0	0
375	Motorcycles, bicycles	1	1	1	0
3914	Silverware & plated ware	0	1	0	7
3942	Dolls	0	0	0	0
3949-R	Sporting & athletic goods	0	0	0	4
396	Costume jewelry	2	0	0	4
Cell III Aggregate		23	7	28	94

(continued)

Table 2.6 (continued)

(d) Cell IV: high export dependence (12.38%)-high import penetration (10.06%)

SIC-Based Code & Title		Protectionist Actions			
		1960-1964	1965-1969	1970-1974	1975-1979
221	Weaving mills, cotton	0	0	0	0
2292	Lace goods	0	0	0	0
237	Fur goods	0	0	0	0
242	Sawmills	0	0	0	0
261	Pulp mills	0	0	0	0
2813	Industrial gases	0	0	0	0
2816-I	Inorganic pigments	4	2	0	7
2819	Industrial inorganic chem	1	3	0	0
2822	Synthetic rubber	0	0	1	1
2833	Medicinals & botanicals	1	0	0	4
2865	Cyclic crude	0	0	0	0
2869-I	Industrial organic chemicals	2	1	6	7
287	Agricultural chemicals	0	0	0	0
311	Leather tanning	0	0	0	0
316	Luggage	0	0	0	0
319	Leather goods nec	0	0	0	0
32113	Laminated glass	0	0	0	0
32115-I	Flat & sheet glass	2	12	9	0
323	Glass products	2	4	5	0
3264	Porcelain electrical supplies	0	0	0	0
3269-I	Pottery products nec	2	0	2	2
342	Cutlery	0	1	3	1
3494	Valves & pipe fittings	0	0	1	0
351	Turbines	0	0	0	0
352	Agricultural machinery	0	0	1	0
354	Metalworking machinery	0	0	0	0
355	Special industrial mach	0	0	1	1
3562	Ball & roller bearings	0	0	1	1

(continued)

Table 2.6 (continued)

SIC-Based Code & Title		Protectionist Actions			
		1960-1964	1965-1969	1970-1974	1975-1979
3563	Air & gas compressors	0	0	0	0
3564	Blowers & fans	0	0	0	1
357	Office & computing equip	2	0	0	1
362	Electrical industrial appar	0	0	0	0
363	Household appliances	0	0	0	0
36514	Phonograph & recorder sys	0	0	1	1
366	Communication equipment	0	0	0	1
367	Electronic components	0	0	4	1
369	Misc electrical equipment	0	0	0	0
37112	Trucks & buses	0	0	0	0
3714	Motor vehicle parts	0	0	0	0
381	Scientific instruments	0	0	0	0
383	Optical instruments	0	0	4	0
385	Ophthalmic goods	0	1	0	0
386	Photographic equipment	0	0	0	2
387	Watches & clocks	1	0	0	0
3911	Jewelry	0	0	1	0
393	Musical instruments	0	4	4	0
3944	Games, toys	0	0	1	0
3949-I	Sporting & athletic goods	1	1	2	5
399	Misc manufactures	2	2	4	0
Cell IV Aggregate		20	31	51	36

* affirmative and negative escape clause and antidumping determinations, and escape clause extensions

I trade with industrialized countries

R trade with the rest-of-the-world

Source: see Table 2.1

An earlier industry petition for escape clause relief against cast iron stoves (targeting South Korea and Taiwan) failed in 1977.

The disaggregated analysis resulted in a net reduction of twenty-five actions in Cell II. In weighted terms, protectionism in the second cell declined 86%. In contrast, thirty-five actions (sixteen escape clause) were added to Cell III. More protected industries are now included in this third cell. In the shortening and cooking oil industry, CVDs were levied against castor oil from Brazil (1976) and oleoresins from Spain (1979). A third motion against oleoresins from India was denied in 1979.

The cellulosic manmade fibers industry (particularly with regard to its trade relations with industrial countries) received protection in five instances in a span of two years. In 1978, rayon staple fiber from Belgium was subjected to antidumping duties. The following year, duties were imposed on the same product; this time from France, Finland, Italy (antidumping) and Sweden (CVD).

In the miscellaneous fabricated wire products industry, CVDs were imposed on iron and steel chains from Italy (1977), Japan, and Spain (1978). In 1978, steel wire strand from Japan became subject to antidumping duties. Other antidumping motions were filed against steel wire products from India (1978) and Canada (1979), but were rejected.

Notable protection was gained by the radio and television industry. In 1977, escape clause measures were taken against imported color televisions. In March 1978, escape clause proceedings resulted in a three-year increase in tariffs on citizen band radio receivers. In both cases Taiwan, Japan and South Korea were the foreign suppliers most affected.

Finally, the analysis produced a net reduction of twenty-two protectionist actions in the crucial fourth cell. Cell IV's weighted protectionism declined 71% from its provisional level. Six more crucial case industries can now be pointed out as manifesting no or minimal protectionist activity despite having "double-digit" import penetration levels: jewelry (34.8%), lace goods (27.3%), industrial inorganic chemicals (16.3%), games and toys (15.2%), industrial gases (11.0%), and ball and roller bearings (10.0%).

We are now in a position to evaluate the aggregate results by cell. The readjusted, reweighted protectionist outcomes for the four cells are presented in Figure 2.3 (directly matching the final sector composition of the 1975-1979 *XD-MP* matrix in Table 2.6). The distribution of protectionism across the four cells and especially the behavior of the crucial fourth cell strongly confirms the export dependence thesis. Low levels of export dependence (2.6% cell average) posed no constraint on the next to highest level of protectionism (192.1) in Cell I. Export dependence and import penetration reinforced each other to produce the two extremes in commercial policy: a liberal Cell II (22.5) and protectionist Cell III (396.6).

Figure 2.3

Final 1975-1979 XD-MP Matrix: Protectionism* by Cell (U.S.)

	LOW EXPORT DEPENDENCE	HIGH EXPORT DEPENDENCE
LOW IMPORT PENETRATION	I 192.1	II 22.5
HIGH IMPORT PENETRATION	III 396.6	IV 50

* weighted affirmative and negative escape clause, antidumping, and countervailing duty determinations, and escape clause extensions; adjusted to the number of cases in cell IV

Source: see Table 2.1

Cell IV manifests the next to lowest level of protectionism (50.0). Despite being significantly more penetrated by imports than Cell I (10.1% vs. 1.9%), Cell IV's protectionism falls well below that of the first cell. In face of substantial threat from imports, export dependence constrained protectionism in Cell IV to a level nearly 75% below that of a cell not faced with such a threat.

Both the third and fourth cells are highly penetrated by imports (13.1% and 10.1%, respectively). Based on this fact alone, one would expect Cell IV's protectionism to be nearly as high as Cell III's. Yet the findings demonstrate quite the contrary; Cell III's protectionism is eight times that of Cell IV. The explanation lies in the high level of export dependence (12.4%) present in the fourth cell but absent (1.8%) in the third.

Cell IV's results come closest to those of the liberal second cell. Despite facing an import penetration level four-and-a-half times greater than Cell II (10.1% vs. 2.2%), Cell IV's protectionism is only two times that of the second cell.

These relationships stand up in a separate examination of the three types of trade action under discussion: escape clause, antidumping, and CVD (see Figure 2.4). The only deviation concerns antidumping cases where Cell IV's protectionism exceeds that of Cell I. Still, the critical comparison between the two highly import penetrated cells demonstrates the marked effect of export dependence. Cell IV's protectionism is exceeded by that of Cell III by a factor of two in terms of antidumping determinations, by a factor of twelve regarding CVD cases, and by 300 when it comes to escape clause decisions.

Figure 2.4

Final 1975-1979 XD-MP Matrix: Escape Claue, Antidumping, and Countervailing Duty Protectionism* by Cell (U.S.)

	Escape Clause	Antidumping	CVD
Cell I	106.8	26.6	58.7
Cell II	0	20.0	2.5
Cell III	235.6	86.4	74.6
Cell IV	0.8	43.0	6.2

* weighted affirmative and negative determinations and escape clause extensions; adjusted to the total number of cases (50) in cell IV

Source: see Table 2.1

The basic findings are closely replicated when considering just those cases where affirmative action or "protection" was taken. In

Figure 2.5, the aggregates of the four cells reflect only those cases that resulted in escape clause relief or the imposition of antidumping or CVD duties (weighted and adjusted to the number of affirmative cases in Cell IV). When it comes to protective measures, Cell IV's level of protection is 61% lower than Cell I, less than twice that of Cell II, and less than one-tenth of Cell III's protection.

Figure 2.5

Final 1975-1979 XD-MP Matrix: Protection* by Cell (U.S.)

	LOW EXPORT DEPENDENCE	HIGH EXPORT DEPENDENCE
LOW IMPORT PENETRATION	I 30.8	II 6.4
HIGH IMPORT PENETRATION	III 133.2	IV 12

* weighted affirmative and negative escape clause, antidumping, and countervailing duty determinations, and escape clause extensions; adjusted to the number of cases in cell IV

Source: see Table 2.1

Figure 2.6

Final 1975-1979 XD-MP Matrix Escape Clause and Antidumping Protectionism* by Cell, 1960-1979 (U.S.)#

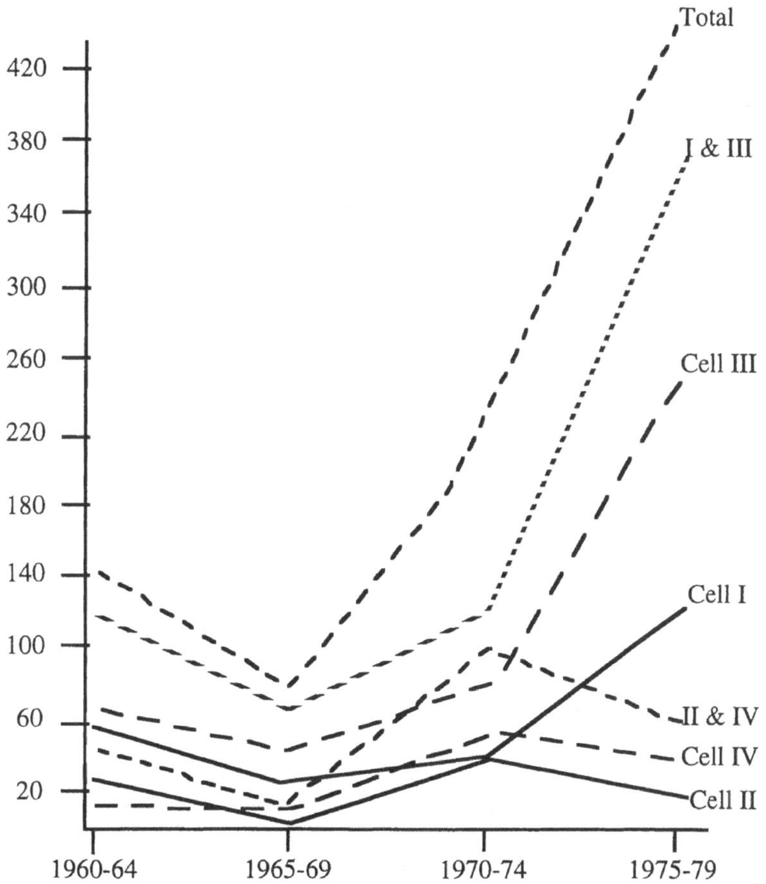

(continued)

Figure 2.6 (continued)

	1960-64	1965-69	1970-79	1975-79
Cell I	45.4	15.4	35.8	109.6
Cell II	17.9	0.1	33.4	16.4
Cell III	57.1	34.1	73.6	264.8
Cell IV	15.1	11.3	51.3	36.0
Cells I & III	102.5	49.5	109.5	374.4
Cells II & IV	33.0	12.1	84.7	52.4
Total	135.6	60.8	194.1	426.8

* weighted affirmative and negative determinations, and escape clause extensions

\# Aggregates trace the protectionism of the XD-MP cells as composed in 1975-1979. They do not represent the protectionist outcomes for the XD-MP cells as composed in the three prior periods. Accordingly, except for the 1975-1979 column, this figure should be read across rows and not down columns.

Sources: see Table 2.1

We can move beyond cross-sectional analysis in testing the export dependence thesis by tracing the behavior of the four cells over time. Figure 2.6 displays the protectionist outcomes for the 1975-1979 cells in terms of all escape clause and antidumping determinations made from 1960 to 1979. (CVD actions are excluded from this longitudinal analysis because official records prior to 1975 are incomplete.)[19] The totals for all sectors show that protectionism in manufacturing was at its lowest point in the 1965-1969 period. This period, of course, encompasses the Kennedy Round of the multilateral trade negotiations (1964-1967), considered to be the zenith of the post-war trade liberalization movement.

The totals also show that the level of protectionism has increased steadily since the 1960s. Many would point to such a trend as evidence of the growing fragility of the liberal trade regime. Yet the cell totals demonstrate that the undeniable increase in U.S. protectionism has been almost entirely attributable to sectors minimally dependent on exports, i.e., Cells I and III. These two cells are responsible for nearly 90% of the increase from the late 1960s.

The two export dependent cells have not, of course, been untainted by protectionism. Yet in the first decade of the new protectionism, protectionism in Cell II declined 52% and dropped 29% in the crucial fourth cell between the early and late 1970s. Over the same period, the level of protectionism shot up 214% and 258% in Cells I and III, respectively. Clearly, the export dependence thesis stands up well in longitudinal analysis.

Just as the export dependence thesis can be tested over time, the analysis can be updated for protectionist activity in 1980. While any given year cannot be indicative of the general influence of export dependence, the record of escape clause motions in 1980 is revealing. Four cases targeting twelve countries were petitioned for the year.

In March 1980, after an ITC recommendation for a three-year increase in duties, the President denied escape clause protection against imported leather wearing apparel. Leather coats and jackets from South Korea, Taiwan, Hong Kong and Argentina would have been subject to the heightened tariffs. The leather garment industry falls in the Cell III miscellaneous apparel sector (SIC 238).

In June, the ITC determined that termination of the orderly marketing arrangements restricting South Korean and Taiwanese color televisions (see following section) would have an adverse economic effect on the domestic television industry. The President accepted the ITC recommendation and extended the import relief for an additional two years. The television industry (SIC 36512) also falls in Cell III.

In November, the mushroom industry gained escape clause protection. While the ITC recommended quantitative restrictions on imports, the President chose to raise tariff duties for a three-year period. Mushrooms from Taiwan, South Korea, and Hong Kong were the major targets of the protective measure. This case involves an industry in the Cell I preserved fruit and vegetables sector (SIC 203).

Finally, the ITC terminated the celebrated case against imported passenger cars in December 1980. The commission determined that automobiles from Japan, Canada and West Germany "are not being imported into the United States in such quantities as to be a substantial cause of serious injury, or the threat thereof, to domestic industries producing articles like or directly competitive with the imported articles."[20]

In Table 2.4 the motor vehicle industry group (SIC 371) was disaggregated at the SIC four- and five-digit level. Two of the three subsectors retained the Cell IV characteristics of the three-digit sector:

trucks and buses (37112) and motor vehicle parts and accessories (3714). The production of passenger cars (37111), however, was found to have Cell III attributes of low dependence (2.5%) and high penetration (18.6%). Hence, in dealing primarily with imported passenger cars, this last escape clause action of 1980 falls in Cell III and not Cell IV.

In sum, all the safeguard cases decided in 1980 concerned low export dependent sectors either in Cell I or III. No activity manifested in the two highly dependent cells, Cell II and the crucial fourth cell.

QUOTAS, VERs, OMAs, TPM, AND MFA

The much celebrated voluntary export restraints and orderly marketing arrangements as well as quotas, the steel trigger price mechanism, and the Multifiber Arrangement can be brought to bear on the analysis of the export dependence thesis. Such arrangements have been imposed on U.S. imports of meat, mushrooms, textiles and apparel, footwear, clothespins, steel, televisions, and automobiles.

Meat. Since 1965, under the Meat Import Act, imports of beef and veal (also mutton and goat meat) have been limited to levels maintaining 1959-1963 ratios of imports to domestic production. Prior to 1979, imports from roughly a dozen Latin American countries plus Australia, New Zealand, and Canada were limited primarily by country-specific quotas. In 1968 the first voluntary export restraint agreements (VERs) for meat were reached with Australia and New Zealand. By 1979, VERs had supplanted the quota system.[21]

Mushrooms. As early as 1964 domestic canners of mushrooms sought escape clause relief. In 1966, the industry requested the President negotiate a VER with Taiwan, the principle supplier of canned mushrooms. The request was denied but Taiwan took steps to restrain its exports to the U.S. in 1968. Canners again sought VERs against South Korea as well as Taiwan in 1972. Negotiations were held, but no agreements reached. A 1975 escape clause petition resulted only in adjustment assistance. In a second case, the ITC recommended tariff-quota restrictions in 1976, but the President rejected any formal import relief. Finally in September 1976, the United States Trade Representative (USTR) held negotiations with Taiwan and South Korea, and received assurances that their exports would be moderated to prevent disruption in the U.S. market. These voluntary restraints remained in effect until late 1979. As discussed above, the VERs were succeeded by formal escape clause protection in 1980. Tariff rates on imported mushrooms were raised from November 1980 through November 1983.[22]

Clothespins. In mid-1978 the clothespin industry filed three "market disruption" cases against the imports of China, Poland, and Rumania. Market disruption cases were essentially the same as escape clause motions except that the former are specifically designed for injurious imports from Communist countries. In August, the ITC found injury from China only and recommended quantitative restrictions. The President denied this relief noting that the ITC had initiated an escape clause investigation concerning clothespins in July. In this latter investigation, the commission again found injury and recommended a five-year global quota of 3,200,000 clothespins per year. In February 1979, the President chose to establish a three-year global quota at 2,000,000 clothespins per year. The arrangement limits a country's imports to the U.S. at 1973-1975 levels — a period prior to the surge of shipments from Taiwan and China.[23]

In each of the three above cases, a Cell I sector has been involved: meat products (SIC 201), preserved fruits and vegetables (203) and miscellaneous wood products (249). The following five restraint arrangements all concern Cell III sectors: nonrubber footwear (314), specialty steels (331), basic steel products (331), television receivers (36512), and passenger cars (37111).

Footwear. The footwear industry first requested, unsuccessfully, escape clause relief in 1970. As a result of a second industry petition, the ITC found in the affirmative and recommended a five-year tariff increase on imported shoes in 1975. The President rejected this proposal and, instead, offered the industry adjustment assistance. In yet a third safeguard case, in February 1977 the Commission recommended a five-year tariff-rate quota system where existing rates of duty would apply to footwear imports within quotas established for each supplier, and increased rates applying to shipments beyond quota limits. This time the President agreed that import relief was necessary and negotiated orderly marketing arrangements (OMAs) with Taiwan and South Korea covering nonrubber footwear shipments from June 1977 to June 1981.[24]

While imports from Taiwan and South Korea were effectively restrained, shipments from other suppliers rose in early 1979. The U.S. expressed concern over a significant increase in Italian shipments and received assurances from EEC officials that such exports to the U.S. would decline substantially in 1980. In 1979, official consultations were also held with the Philippines, Singapore, Brazil and Hong Kong. In each case, exports to the U.S. declined notably after prior surges.[25] In June 1981, despite an ITC recommendation for their extension, the President permitted the two formal OMAs with Taiwan and South Korea to expire.

Specialty Steels. Between 1969 and 1971, Japanese and European Community producers agreed to VERs limiting shipments of steel mill products to the U.S. In early 1972, the VERs were extended through 1974 for Britain as well as the EC and Japan. Sublimits for specialty steels were established at this time. In mid-1975 the specialty steel industry petitioned for an escape clause investigation of imported stainless and alloy tool steels. The ITC recommended a five-year quota system and in March 1976 the President directed the USTR to negotiate OMAs with key specialty steel suppliers. In June, an OMA with Japan was reached and import quotas were imposed on all other suppliers from June 1976 through June 1979. Besides Japan, the quotas targeted the EC, Canada, and Sweden. In November 1978, the industry requested the renewal of the quotas. The ITC split evenly on this issue, but the President extended the restrictions until February 1980.[26]

Steel Trigger Price Mechanism. In a move precipitating a major trade crisis in the fall of 1977, domestic steel concerns filed no less than eighteen antidumping petitions (and threatened more) against Japanese, European and LDC steel producers. To resolve the crisis and gain the withdrawal of the petitions, an interagency task force recommended and the President approved the establishment of the steel trigger price mechanism (TPM) in late 1978. Based on the costs of production for the most efficient steel producer (Japan), trigger prices were set for most basic steel mill products. Where imports were shipped below trigger prices, the Department of Commerce initiates "fast track" antidumping cases.[27] From May 1978 (when it became operational) through October 1981, twelve antidumping cases have been initiated by the TPM: carbon steel plate from Taiwan, Poland, and Spain (October 1978); carbon steel plate from Belgium, France, West Germany, Italy, and Britain (December 1978); steel wire nails from South Korea (April 1979); and steel wire nails from Japan, Yugoslavia, and South Korea (July 1981).

Televisions. After an industry safeguard petition, in March 1977 the ITC recommended a five-year tariff increase on imported color televisions. In response, the President negotiated an OMA with Japan limiting that country's exports from July 1977 to June 1980. When television shipments from Taiwan and South Korea increased sharply after the Japanese OMA went into effect, OMAs were also negotiated with these two other countries effective February 1979 through June 1980. Color television imports from Japan dropped 80% in terms of quantity from 1976 to 1979. Between 1978 and 1979, Taiwanese and South Korean shipments declined 41% and 28%, respectively. In December 1979, the television industry sought the extension of all three OMAs. In June 1980, the President accepted the ITC's recommendation to renew only the Taiwanese and South Korean restraints through June 1982.[28]

Automobiles. In June 1980, the automobile industry sparked a nearly year-long international trade controversy by petitioning for escape clause relief from imports. With a negative determination in December, the ITC terminated the case but by no means the controversy. As a result of pressure from both within and outside the administration, the President gained a three-year voluntary export restraint from the Japanese in April 1981. For the first year, Japan agreed to limit shipments to the U.S. to 1,680,000 passenger cars — an 8% reduction from 1980 levels. The same quantitative restriction applies for the second agreement year plus the Japanese are allowed 16.5% of any expansion in the U.S. automobile market. Shipments are to be monitored and surges guarded against in the third year.[29]

Up to this point, none of the above import restrictions run counter to the export dependence thesis. All eight arrangements deal with low export dependent industries — those falling either in Cell I or III. The argument also holds up in an examination of the Multifiber Arrangement covering textile and apparel imports.

Multifiber Arrangement. In the early 1950s, the textile industry became increasingly concerned over import penetration by Japanese cotton textiles and began requesting escape clause relief and import quotas. In late 1955, Japan initiated internal restraints on textile exports to the U.S. and in January 1957 entered into a five-year VER with the U.S. As a result of Japan's voluntary restraints, cotton textile imports dramatically rose from other sources — notably, Hong Kong. In the late '50s unsuccessful overtures were made to Hong Kong to implement controls similar to Japan's.

The U.S. was the first to bring the textile trade issue to the GATT in late 1959, and the President's May 1961 textile program included a call for multilateral consultations on the matter. In July, GATT members agreed to the Short-Term Arrangement Regarding International Trade in Cotton Textiles (STA). The STA provided for bilateral agreements (similar to OMAs) and unilateral quotas to restrain imports. Restraints from October 1961 to September 1962 were set at levels not less than export levels during 1960-1961. Cotton textile restrictions were extended through September 1973 with the adoption of the Long-Term Arrangement (LTA) in February 1962, and its renewal in 1967 and 1970. The STA, LTA as well as the Multifiber Arrangement have had the dual objectives of gaining "increased access to the industrial national markets for exports of developing countries" and "ensuring the orderly development of this trade and avoidance of disruptive effects in industrial markets."[30]

The STA and LTA applied only to cotton textiles, and through the 1960s textile producers increased production of wool and manmade fiber exports. In response, in the fall of 1971 the U.S. negotiated its first bilateral agreements limiting imports of manmade fiber and wool textiles. In January 1974 textile exporters and importers agreed to restraints for all fibers under the Multifiber Arrangement (MFA). The MFA was renewed in 1977 and 1981.

Prior to the 1977 extension of the MFA, textile producers were guaranteed at least 6% annual growth rates in exports. By 1977, however, the U.S. and European textile importers began tightening their import controls primarily by negotiating bilateral agreements restricting growth rates well below the 6% level. In 1978, the MFA was amended to legitimate such "jointly agreed reasonable departures" from arrangement provisions. In March 1979, the President initiated a comprehensive textile industry program which included the strict monitoring of bilateral agreements to control import surges and reduce the flexibility with which limits were adhered to. In addition, the terms of agreements with Hong Kong, Taiwan and South Korea were renegotiated in 1979. The heightened restrictions were also evident in the unilateral quotas imposed on China in 1979, the renewal of the Sri Lanka agreement in 1980, and the 1980 China bilateral.[31]

As of late 1980, the U.S. had textile restraint agreements with twenty-two countries. In addition, "consultation" agreements with ten other countries provided for surveillance of import levels and potential negotiated or unilateral restraints. Limits are imposed on 104 MFA product categories that cover virtually all SIC textile and apparel product codes and cut across the four cells of the 1975-1979 *XD-MP* matrix. The question arises as to whether any segment of textile and apparel imports has borne the brunt of protection under U.S. MFA restraints. In particular, are there notable differences in restrictions on low versus highly export dependent sectors?

These issues can be explored by examining the structure of protection in U.S. bilaterals. In any given year, a large majority of product category limits are not filled. Well over half of U.S. textile and apparel imports do not fill 85% of their agreement levels (see aggregate totals in Figure 2.7a). The extent to which negotiated restraint levels actually have the effect of limiting imports varies from product to product.

The Commerce Department closely monitors those product categories whose imports fill 85% or more of their agreed restraint levels. We can consider such shipments as coming under "effective restraint." We can now consider whether levels of effective restraint vary across the *XD-MP* cells. As shown in Figure 2.7a, in the late 1970s a higher proportion of imports in the two minimally export dependent cells did, indeed, come under effective restraint compared to imports in the highly export dependent cells. In 1976, over 64% of textile apparel imports falling in Cell III filled at least 85% of their agreed limits; close to 33% of Cell I imports were subject to effective restraints. In contrast, only 2% of textile imports in Cell II and under 28% of Cell IV imports came under effective restraint.

Longitudinal analysis reveals significant changes in the structure of U.S. MFA protection. In 1969, Cell IV imports faced as much effective restraint as Cell III shipments. Agreement limits were filled at the 85% level or beyond by 50% and 57% of Cell IV and Cell III imports, respectively. Indeed, prior to the 1970s, Cell IV imports, notably cotton weaving mill products, were the primary targets of U.S. textile restraints. In 1969, over 50% of all effective restraints targeted Cell IV products (see Figure 2.7b). In fact, two of the three products limited in the original 1957 VER with Japan were cotton weaving products: velveteens and ginghams (the third target being women's and children's cotton blouses).

Figure 2.7

Structure of U.S. Protection in MFA Bilateral Restraint Agreements

(a) Share of Textile & Apparel Imports Under Effective Restraint* by 1975-1979 Cell

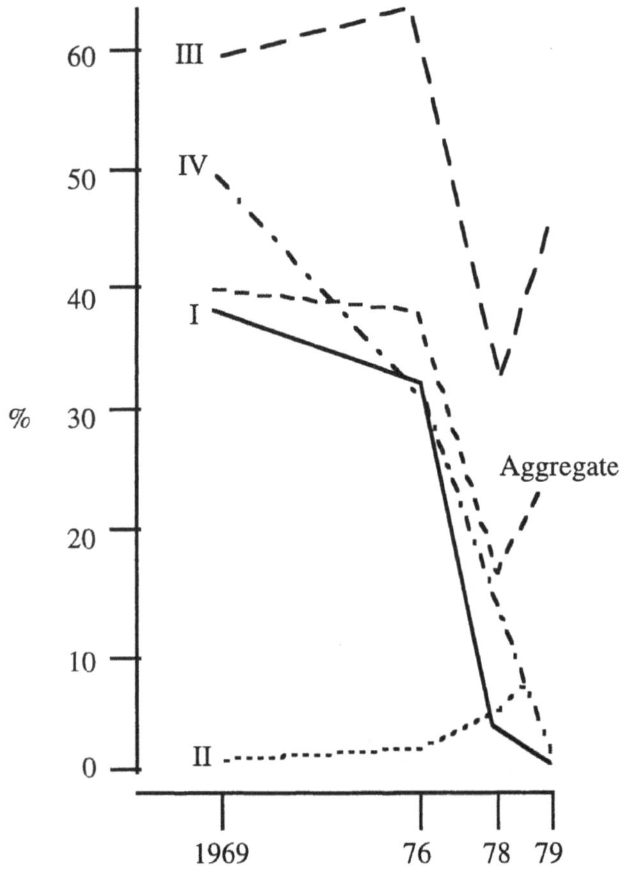

* 85% of product category agreed limit filled

(continued)

Figure 2.7 (continued)

(b) Distribution of Effective Restraints* by Cell

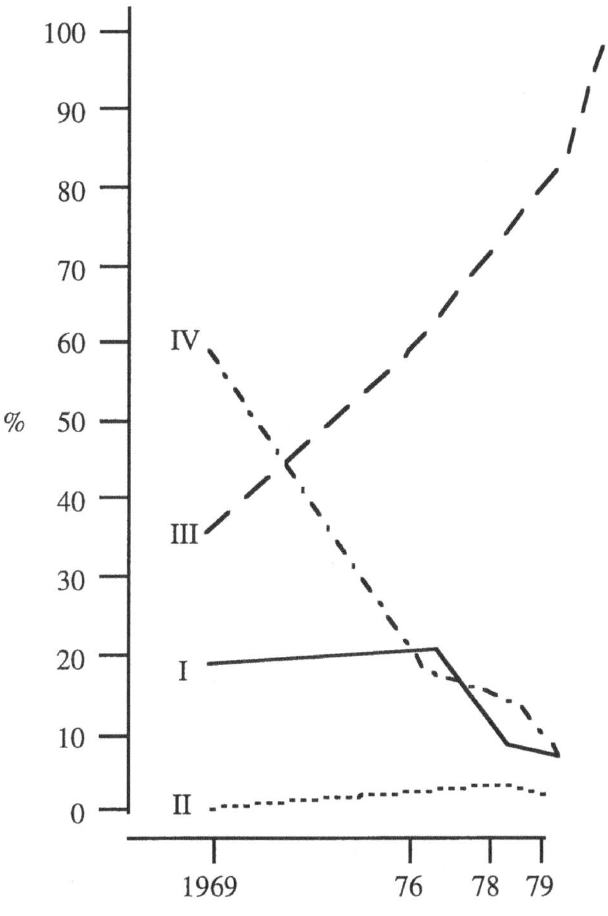

* 85% of product category agreed limit filled

(continued)

Figure 2.7 (continued)

Sources: U.S., Bureau of the Census, *U.S. General Imports of Cotton Manufactures: Country & Geneva Agreement Category,* FT 130
U.S., Congress, House of Representatives, Committee on Ways and Means, Subcommittee on Trade, *Background Material on the Multifiber Arrangement,* Hearing, 96th Congress, 1st Session, April 2, 1979
U.S., Department of State, *Treaties & Other International Acts Series*
____, *Treaties in Force*
____, *United States Treaties & Other International Agreements*
U.S., Department of Treasury, *Customs Bulletin & Decisions*
U.S., International Trade Commission, *The History & Current Status of the Multifiber Arrangement* (USITC Publication 850, January 1978)
____, *The Multifiber Arrangement:* 1973-1980, 3 v. (USITC Publication 1131, March 1981)

This protection against cotton weaving mill products reflects the fact that in the 1960s this industry was not, in contrast to the 1970s, highly dependent on exports. In 1962 and 1967, levels of export dependence were only 4.53% and 4.75% respectively; compared to 7.96% and 10.65% in 1972 and 1977, respectively.[32] Prior to the seventies, the industry faced little policy constraint from export dependence. However, once the industry became increasingly dependent on exports, cotton mill protection became circumscribed. Whereas 50% of Cell IV imports were effectively restrained and 51% of effective restraints targeted Cell IV imports in 1969, shipments competing with crucial fourth cell industries were totally unshackled from such limits by 1979. Instead, U.S. MFA restraints have transformed to target apparel imports. While some restrictions on yarn and thread have been loosened, quotas on many apparel items have been frozen since the crack-down on U.S. bilaterals in 1977.

In sum, since the 1970s U.S. MFA textile agreements have primarily targeted imports competing with minimally export dependent industries — particularly clothing industries in Cell III. Consequently, none of the nine protectionist arrangements discussed in this section contradict the export dependence thesis. Three arrangements concern Cell I sections: meat VERs, canned mushroom VERs, the clothespin global quota. And six involve Cell III industries nonrubber footwear OMAs, specialty steel OMAs, the steel trigger price mechanism, color television OMAs, the automobile VER, and the MFA.

ALTERNATIVES TO THE EXPORT DEPENDENCE THESIS

Before concluding, we can briefly examine the relative explanatory power of the export dependence thesis, and potential spuriousness of the relationship between export dependence and protectionism. Three major alternative factors or variables — economic performance, foreign direct investment, and technology intensity — may either be responsible for the results of the above aggregate data analysis, and/or serve as more powerful predictors of industry protectionism than levels of export dependence.

In particular, export dependent industries may manifest low levels of protectionism not because of their dependence upon exports, but because these industries may also happen to be technologically sophisticated, high growth sectors engaged in transnational investment activity. The latter characteristics, may be the determining variables driving the results of the aggregate data analysis, rendering the relationship between export dependence and industry trade policy positions as spurious. A related possibility is that the economic performance, technological sophistication, and foreign direct investment activity of the 129 industrial sectors may be better predictors of their trade policy positions than their levels of export dependence. Both possibilities are examined in this section.

The role that economic growth may play in dampening protectionism was briefly noted in Chapter I. If the sectors falling into the highly export dependent cells also happen to be high-growth industries, then industrial economic performance and not export dependence may account for the low levels of protectionism found in Cells II and IV.

Liberal trade policy views may also be explained in terms of the technology intensity of an industry. Technological innovation and leadership could provide an industry the confidence to compete at home and abroad, and, therefore, offer "technological" alternatives to "political" strategies of protectionism.

Technology may also provide the backdrop for the export positions of export dependent industries. Technological advantages may help establish the comparative advantage of an industry in world markets. If export dependent industries manifest low levels of protectionism, and if industries are export oriented because of technological sophistication, then link between export dependence and liberal trade policy views may be nothing more than a spurious correlation.

That link may also be spurious because industry export positions may be derivatives of the foreign investment activities. There is an integral and complex relationship between foreign direct investment (FDI) and exports. There is no doubt that TNEs play a prominent role in U.S. exports; in 1970 over 20% of U.S. exports represented sales of U.S. parent companies to their foreign subsidiaries.[33] When the issue turns, however, to the positive or negative effect of foreign direct investment on exports, the picture becomes less clear.[34] On the one hand, in jumping tariff walls and directly serving foreign markets, foreign direct investment displaces the export of finished goods. As subsidiaries move from assembly and packing operations to manufacturing, exports may also be supplanted.

On the other hand, FDI enhances "associated exports." The establishment of subsidiaries requires initial exports of capital equipment. More generally, foreign operations may require the transfer of machinery, parts and components, raw materials, and certain finished supplies. Exports are enhanced to the extent foreign operations are "market expanding investments" that engage in market research, wholesale distribution, service and repair, and parts replacement. While small relative to subsidiary sales in and from the host country, "associated exports" from the parent company may be significantly larger than would be the case in the absence of the foreign subsidiary.[35]

Just as FDI may underlie industry export positions, foreign investment activity may motivate industry liberal trade policy views.[36] TNEs have often become advocates of free trade and payments since it is in their interests to lower, like transportation costs, the burdens of foreign duties and taxes, currency controls, government regulations, and administrative requirements.

The following operational indicators for foreign direct investment and technology intensity as well as five indicators for economic performance are utilized to examine their interrelationships with export dependence and protectionism:

> *Foreign Direct Investment proxy:* ratio of dividends of foreign corporations plus foreign tax credit to total assets, 1976

Research and Development proxy: ratio of engineers and scientists to total employment, 1977

Growth: 1972-1978 derived annual average rate of change in value-added

Profit-Sales ratio: ratio of net income minus income tax to total receipts, 1976

Profit-Equity ratio: ratio of net income minus income tax to net worth, 1976

Labor Productivity: 1972-1978 derived annual average rate of change in ratio of production worker hours to value-added

To check on the potential spuriousness between export dependence and protectionism, the 129 SIC-based three-digit industry groups were scored on each of the above indicators and then correlated against their levels of export dependence. Levels of export dependence were also correlated against import penetration ratios. Table 2.7 displays the simple Pearson product-moment correlation coefficients. The striking feature of the results is the absence of any strong correlations between this study's independent variable and the seven "alternate" variables (only three relationships are even statistically significant at the 0.05 level). Except for the R&D proxy (with the least weak coefficient value of 0.48), the alternate variables are virtually randomly associated with export dependence levels. The weak relationship (0.16) between foreign investment and export dependence confirms the earlier findings of Bergsten, Horst and Moran, among others.[37]

Table 2.7

Correlation of Export Dependence with Alternate Variables (U.S.)

	MP	FDI	GR	P/S	P/E	R&D	LP
XD	-0.0226[a]	0.1648	0.1845	0.1927	0.0474	0.4793	-0.0715
	(*)[b]	(*)	(0.0364)	(0.0286)	(*)	(0.0001)	(*)

XD export dependence: ratio of exports to shipments plus imports; 1977 adjusted by derived annual average rate of change over 1972

MP import penetration: ratio of imports to shipments plus imports; 1977 adjusted by derived annual average rate of change over 1972

FDI foreign direct investment proxy: ratio of dividends of foreign corporations plus foreign tax credit to total assets, 1976

GR growth: 1972-1978 derived annual average rate of change in value-added (1972 prices)

P/S profit-sales ratio: ratio of net income minus income tax to total receipts, 1976

P/E profit-equity ratio: ratio of net income minus income tax to net worth, 1976

R&D research and development proxy: ratio of engineers and scientists to total employment, 1977

LP labor productivity: 1972-1978 derived annual average rate of change in ratio of production worker hour to value-added (1972 prices)

(continued)

Table 2.7 (continued)

[a] Pearson's r
[b] significance
* not significant at the 0.05 level

Sources: U.S., Bureau of Economic Analysis (SIC two-digit price deflators)
U.S., Bureau of Labor Statistics, *Occupational Employment Statistics Survey of Manufacturing* (unpublished)
U.S., Bureau of the Census, *Annual Survey of Manufactures: Statistics for Industry Groups Industries*
U.S., Internal Revenue Service, *Corporation Sourcebook of Statistics of Income* (unpublished)

These findings can be fleshed out by examining the performance of the crucial fourth cell sectors on the alternate variables. Their nominal scores on the variables, based on their location above or below weighted means for each variable, are presented in Table 2.8.[38] Of the 40 highly dependent Cell IV sectors that could be individually scored on the alternate variables, only 22 are also rank highly on the foreign direct investment proxy, and 21 highly on profit-sales ratio and the R&D proxy. The overlap is even smaller for the three other variables. Significantly, only 14 of the 40 sectors performed above the mean for economic growth.

Table 2.8

Cell IV Sector Performance on Alternate Variables (U.S.)

SIC-Based Code	Title	FDI	GR	P/S	P/E	R&D	LP
221	Cotton weaving	L	L	L	L	L	H
2292	Lace goods	L	L	L	L	L	H
237	Fur goods	L	L	L	H	L	L
242	Sawmills	L	H	H	H	L	H
261	Pulp mills	H	H	H	L	H	H
2813	Industrial gases						
2816-I	Inorganic pigments	H	H	H	L	H	L
2819	Industrial inorganic chemicals						
2822	Synthetic rubber	H	L	H	L	H	L
2833	Medicinals	H	L	H	H	H	L
2865	Cyclic crude						
2869-I	Industrial organic chemicals	H	H	H	L	H	H

(continued)

Table 2.8 (continued)

SIC-Based Code & Title		FDI	GR	P/S	P/E	R&D	LP
287	Agricultural chemicals	H	H	L	L	H	H
311	Leather tanning	L	L	L	L	L	H
316	Luggage	L	H	L	L	L	H
319	Leather goods nec	L	H	L	L	L	L
32113	Laminated glass						
32115-I	Flat glass	H	L	H	L	L	L
323	Glass Products	H	H	H	L	L	L
3264	Porcelain electrical supplies						
3269-I	Pottery products nec	H	L	L	L	L	H
342	Cutlery	H	H	H	H	L	L
3494	Valves & pipe fittings	L	H	H	H	L	H
351	Turbines	L	L	L	L	H	L
352	Agricultural machinery	L	L	H	H	H	L
354	Metalworking machinery	L	L	L	L	H	L
355	Special industrial machinery	H	L	H	H	H	L
3562	Ball & roller bearings						
3563	Air & gas compressors	H	H	H	L	H	L
3564	Blowers & fans						

(continued)

Table 2.8 (continued)

SIC-Based Code & Title		FDI	GR	P/S	P/E	R&D	LP
357	Office & computing machines	H	H	H	H	H	L
362	Electrical industrial apparatus	L	L	H	H	H	L
363	Household appliances	H	L	H	H	L	L
36514	Phonograph & recorder systems	H	L	L	L	H	L
366	Communication equipment	H	L	L	L	H	L
367	Electronic components	L	L	L	L	H	L
369	Misc electrical equipment	L	H	H	H	H	L
37112	Trucks & buses						
3714	Motor vehicle parts	H	L	H	H	H	L
381	Scientific instruments	H	L	L	L	H	L
383	Optical instruments & lenses	H	H	H	H	H	L
385	Opthalmic goods	H	L	H	H	L	L
386	Photographic equipment	H	L	H	H	H	L
387	Watches & clocks	H	L	L	L	L	L
3911	Jewelry	L	L	L	H	L	L
393	Musical instruments	L	L	L	H	L	L
3944	Games, toys						
3949-I	Sporting & athletic goods	L	L	L	H	L	H

(continued)

Table 2.8 (continued)

SIC-Based Code & Title	FDI	GR	P/S	P/E	R&D	LP
399 Misc Manufactures	L	L	L	H	L	H
# High (N=40)	22	14	21	19	21	12
Total Manufacturing (N=120)						
Weighted mean*	0.81	7.50	3.31	9.61	2.09	6.92
Unweighted mean	0.75	6.77	3.05	8.84	2.31	6.67
Aggregate national average	1.44	8.46	3.27	9.64	2.99	7.80

H high: at or above weighted mean
L low: below weighted mean
* used for sector classification

Sources: see Table 2.7

To compare the relative superiority of these alternative variables versus export dependence in explaining levels of protectionism, levels of escape clause, antidumping, and CVD protectionism were determined for the industrial sectors falling in the top two quartiles (i.e., above the weighted means) for export dependence and the seven alternate variables (the lowest two quartiles were used in the case of import penetration). Table 2.9 shows the percentage of 1975-1979 manufacturing protectionism found in the top quartiles for each variable.

Table 2.9

Share of 1975-1979 Manufacturing Protectionism* in Top Two Quartiles# of Export Dependence and Alternate Variables

Export Dependence:	9.37%
Import Penetration:	35.39
Foreign Direct Investment:	30.91
Growth:	70.65
Profit-Sales ratio:	29.00
Profit-Equity ratio:	24.65
Research & Development:	21.76
Labor Productivity:	65.95

* weighted affirmative and negative escape clause, antidumping, and countervailing duty determination, and escape clause extensions

sectors at or above weighted mean for each variable

Sources: see Tables 2.1 and 2.7

The results bear out the relative superiority of the export dependence thesis. While constituting one-half of the manufacturing economy, highly export dependent sectors were responsible for only 9% of all manufacturing protectionism. The corresponding shares of protectionist behavior found in high technology, profitable, and transnational enterprises were 21%, 25-29%, and 31%, respectively. In the cases of labor productivity and growth, well over half of all protectionism can actually be found in sectors performing highly on the two variables.

Without any pretense regarding their representativeness, a whole series of highly protectionist sectors can be pointed out as performing well on the alternate variables (see Table 2.10). Chemical and metal fabricating industries (SIC 28 and 34) stand out as protectionist sectors scoring high on up to five of the six alternate variables (excluding import penetration).

This is not to say that the alternate variables do not have an impact on protectionist behavior. Anticipated negative relationships were borne out in the five cases where less than half of manufacturing protectionism manifested in the respective top quartiles. The keynote to conclude this section with is that export dependence has a sufficiently autonomous if not superior impact on levels of protectionism to warrant the attention given it in this study.

Table 2.10

Protectionist Sectors* Scoring High# on Alternate Variables

SIC-Based Code	& Title	FDI	GR	P/S	P/E	R&D	LP
202	Dairy products	X			X		
203	Preserved fruits & vegatables	X	X				X
206	Sugar products		X				X
233	Women's outerwear				X		
234	Women's undergarments				X		
235	Hats, caps				X		
236	Misc outerwear				X		
238	Misc apparel				X		
249	Misc wood products		X				X
281	Industrial inorganic chem	X	X	X		X	
282	Plastics materials	X		X		X	
283	Drugs	X		X	X	X	
286	Industrial organic chem	X	X	X		X	X
307	Misc plastics products		X				
314	Footwear, except rubber				X		X
317	Handbags		X				X

(continued)

Table 2.10 (continued)

SIC-Based Code & Title		FDI	GR	P/S	P/E	R&D	LP
321	Flat glass	X		X			
331	Basic steel products		X				X
345	Screw machine products	X	X	X	X		
349	Misc fabricated metal prod		X	X	X		X
365	Radio & television equipment	X				X	
371	Motor vehicles	X		X	X	X	
391	Jewelry, silverware				X		
394	Toys & sporting goods				X		

* sectors with five or more protectionist actions (escape clause, antidumping, and CVD) and/or subject to VERs, OMAs, quotas, the steel trigger price mechanism, or the Mulitifiber Arrangement

\# sector at or above weighted mean for designated variables

Sources: see Tables 2.1 & 2.7

SUMMARY

The findings of this chapter can be concisely summarized. The export dependence thesis was confirmed in an aggregate data analysis of 129 U.S. industrial sectors. In particular, the hypothesized negative relationship between export dependence and protectionism has stood up under:

(a) cross-sectional analysis of 1975-1979 escape clause, antidumping, and CVD protectionism;

(b) disaggregated analyses of escape clause, antidumping, and CVD cases;

(c) consideration of only those cases that resulted in protective action;

(d) longitudinal analysis of escape clause, antidumping, and CVD protectionism from 1960 to 1979;

(e) an updated examination of 1980 escape clause cases; and

(f) an examination of voluntary export restrictions on meat, mushrooms and automobiles; orderly marketing arrangements for footwear, specialty steel, and televisions; quotas for clothespins; the steel trigger price mechanism; and the Multifiber Arrangement.

Relatively low levels of protectionism were found even in those cases where highly export dependent industries were also highly penetrated by imports. Levels of export dependence have also been found to be significantly independent of and superior in explanatory power to foreign direct investment, industrial growth, profit ratios, research and development, and labor productivity.

Attention now turns to the elaboration of the export dependence thesis in the case study analysis of a selected "crucial case" industry.

NOTES

1 The 129 sectors are based on a cross-indexation of Bureau of the Census SIC codes for export products, import products, and product shipments at the three-digit level. Sources: U.S., Bureau of the Census, *Numerical List of Manufactured Products*, October 1978; and Table 2, "Principle Differences Between the SIC-Based Output and Export/Import Codes" in U.S., Bureau of the Census, *U.S. Exports: SIC-Based Products by World Area*, FT 610 and *U.S. Imports: SIC-Based Products by World Area*, FT 210. A full specification of the SIC coverage of each of the 129 sectors can be obtained from the author.

2 These operational considerations are offered by Clark A. Murdock, "Economic Factors as Objects of Security" in *Economic Issues and National Security*, Klaus Knorr and Frank N. Trager, eds. (Lawrence, Kansas: The Regents Press of Kansas, 1977); and Kenneth N. Waltz, "The Myth of National Interdependence" in *The International Corporation*, Charles P. Kindleberger, ed. (Cambridge: MIT Press, 1970).

3 Ratio from Alan Hewer, "Manufacturing Industry in the Seventies: An Assessment of Import Penetration and Export Performance," *Economic Trends*, n. 320 (Great Britain, Central Statistical Office, June 1980); J.D. Wells ad J.C. Imber, "The Home and Export Performance of United Kingdom Industries," *Economic Trends*, n. 287 (Great Britain, Central Statistical Office, August 1977).

4 Richard Blackhurst, Nicolas Marian and Jan Tumlir, *Trade Liberalization, Protectionism and Interdependence* (GATT Studies in International Trade, n. 5, November 1977), pp. 7-18. See also Wells and Imber, "The Home and Export Performance of United Kingdom Industries," pp. 79-80.

5 Blackhurst, *et al,*, *Trade Liberalization, Protectionism and Interdependence*, p. 15.

6 Ratio from Hewer, "Manufacturing Industry in the Seventies;" Wells and Imber, "The Home and Export Performance of United Kingdom Industries."

7 For some of the voluminous literature on the new protectionism see Bela Balassa, "The 'New Protectionism' and the International Economy," *Journal of World Trade Law*, v. 12, n. 5 (September-October 1978); Richard Blackhurst, *et al, Trade Liberalization, Protectionism, and Interdependence*; V. Cable, "British Protectionism and LDC Imports," *Overseas Development Institute Review*, v. 2 (1977); "Britain, the New Protectionism, and Trade with

the Newly Industrializing Countries," *International Affairs*, v. 55, n. 1 (January 1979); Gerard Curzon, "Neo-protectionism, the MFA and the European Community," *World Economy*, v. 4, n. 3 (September 1981); J.M. Finger, " The Industry-Country Incidence of 'Less than Fair Value' Cases in U.S. Import Trade," *Quarterly Review of Economics and Business*, v. 21, n. 2 (Summer 1981); Lawrence G. Franko, "Current Trends in Protectionism in Industrialized Countries: Focus on Western Europe," in *Protectionism or Industrial Adjustment?*, G.K. Heilleiner, *et al.*, ed. (Paris: Atlantic Institute for International Affairs, April 1980); Linda M. Gard and James Riedel, "Safeguard Protection of Industry in Developing Countries," *Weltwirtschaftliches Archiv*, b. 116, h. 3 (1980); G.K. Heilleiner, "The New Industrial Protectionism and the Developing Countries," in *Protectionism or Industrial Adjustment?*; Brian Hindley, "VERs and GATT's Main Escape Clause," *World Economy*, v. 3, n. 3 (November 1980); Melvin Krauss, *The New Protectionism: The Welfare State and International Trade* (New York: New York University Press, 1978); Morris E. Morkre and David G. Tarr, *The Effects of Restrictions on United States Imports: Five Case Studies and Theory*, Staff Report of the Bureau of Economics to the Federal Trade Commission (Washington: Government Printing Office, June 1980); Edward L. Morse, "The New Economic Nationalism and the Coordination of Economic Policies," in *The New Nationalism: Implications for Transatlantic Relations*, Werner Link and Werner J. Feld, eds. (New York: Pergamon Press, 1979); Tracy Murray, *et al.*, "Alternative Forms of Protection Against Market Disruption," *Kyklos*, v. 31, f. 4 (1978); Bahram Nowzad, *The Rise in Protectionism* (Washington: International Monetary Fund, 1978); Gary P. Sampson, "Contemporary Protectionism and the Exports of Developing Countries," *World Development*, v. 8 (Fall 1979); Susan Strange, "The Management of Surplus Capacity: Or How Does Theory Stand Up to Protectionism 1970s Style?", *International Organization*, v. 33, n. 3 (Summer 1979); Ingo Walter and Jae W. Chung, "The Pattern of Non-Tariff Obstacles to International Market Access," *Weltwirtschaftliches Archiv*, b. 108, h. 1 (1972).

8 Nowzad's much cited study of the new protectionism (*The Rise in Protectionism*) also limits itself to restrictions affecting imports (p. 8).

9 Empirical studies have demonstrated that various forms of non-tariff barriers and tariff restrictions are highly correlated. See Finger, "The Industry-Country Incidence"; Edward John Ray, "The Determinants of Tariff and Non-Tariff Trade Restrictions in the United States," *Journal of Political Economy*, v. 89, n. 1 (February 1981); Walter and Chung, "The Pattern of Non-Tariff Obstacles."

10 Section 201 of the Trade Act of 1974.

11 Section 201 of the Antidumping Act of 1921.

[12] With the establishment of the steel trigger price mechanism in 1978, antidumping cases involving steel products could be initiated by the Commerce Department.

[13] Section 303 of the Tariff Act of 1930.

[14] Blackhurst, *et al.*, *Trade Liberalization, Protectionism and Interdependence*, p. 44.

[15] For similar conclusions see Bela Balassa, "The 'New Protectionism' and the International Economy;" Finger, "Industry-Country Incidence of 'Less than Fair Value' Cases in U.S. Import Trade," pp. 262-65; and Nowzad, *The Rise in Protectionism*, p. 9.

[16] Major suppliers were determined by examining ITC investigation reports and/or as specified in the Trade Action Monitoring System issued by the Office of the United States Trade Representative.

[17] The three-digit sector weighted means of 4.76% for export dependence 5.25% for import penetration are used for the cell classification of the four- and five-digit subsectors.

[18] Export dependence and import penetration levels for the 129 sectors disaggregated by their 1977 trade with industrial countries versus the rest of the world are available from the author.

[19] Records of affirmative CVD determinations are apparently complete. However, prior to 1975 records of negative decisions are virtually nonexistent. This bureaucratic anomaly was recognized, in apparent dismay, in an in-house Treasury Department memo: "... negative decisions in countervailing duty cases have been made on a very informal basis and have rarely been published." ("Countervailing Duties Categories of Bounty-or-Grant Issues," January 30, 1967.) Published accounts of negative determinations prior to 1975 have been found for only three cases (one of which was acknowledged in the Treasury memo).

[20] U.S., International Trade Commission, *Certain Motor Vehicles and Certain Chassis and Bodies Therefor: Report to the President on Investigation No. TA-201-44 Under Section 201 of the Trade Act of 1974* (USITC Publication 1110, December 1980), p. 1.

[21] U.S., International Trade Commission, *Live Cattle and Certain Edible Meat Products of Cattle: Report to the President on Investigation No. TA-201-25 Under Section 201 of the Trade Act of 1974* (USITC Publication 834, September 1977).

[22] U.S., International Trade Commission, *Mushrooms: Report to the President on Investigation No. TA-201-17 Under Section 291 of the Trade Act of 1974* (USITC Publication 798, January 1977); *Mushrooms: Report to the President on Investigation No. TA-201-43 Under Section 201 of the Trade Act of 1974* (USITC Publication 1089, August 1980).

[23] U.S., International Trade Commission, *Clothespins: Report to the President on Investigation No. TA-201-36 Under Section 291 of the Trade Act of 1974* (USITC Publication 933, December 1978); U.S. President, *Annual Report of the President of the United States on the Trade Agreements Program, 1979*, p. 106.

[24] U.S., International Trade Commission, *Footwear: Report to the President on Investigation No. TA-201-7 Under Section 201 of the Trade Act of 1974* (USITC Publication 758, February 1976); *Footwear: Report to the President on Investigation No. TA-201-18 Under Section 201 of the Trade Act of 1974.*

[25] U.S., President, *Annual Report of the President of the United States on the Trade Agreements Program, 1979*, p. 107.

[26] U.S., International Trade Commission, *Stainless Steel and Alloy Tool Steel: Report to the President on Investigation No. TA-201-5 Under Section 201 of the Trade Act of 1974* (USITC Publication 756, January 1976); *Stainless Steel and Alloy Tool Steel: Report to the President on Investigation No. TA-203-3 Under Section 203(i)(2) of the Trade Act of 1974* (USITC Publication 838, October 1977; *Stainless Steel and Alloy Tool Steel: Report to the President on Investigation No. TA-203-5 Under Section 203 of the Trade Act of 1974* (USITC Publication 968, April 1979); U.S., President, *Annual Report of the President of the United States on the Trade Agreements Program, 1979*, pp. 103-104.

[27] U.S., President, *Annual Report of the President of the United States on the Trade Agreements Program, 1979*, pp. 102-103.

[28] U.S., International Trade Commission, *Television Receivers, Color and Monochrome, Assembled or Not Assembled, Finished or Not Finished, and Subassemblies Thereof: Report to the President on Investigation No. TA-201-19 Under Section 201(b) of the Trade Act of 1974* (USITC Publication 808, March 1977); *Color Televisions Receivers and Subassemblies Thereof: Report to the President on No. TA-203-b Under Section 203 of the Trade Act of 1974* (USITC Publication 1068, May 1980).

29 *Economist*, May 9, 1981; U.S., International Trade Commission, *Certain Motor Vehicles*.

30 U.S., International Trade Commission, *The History and Current Status of the Multifiber Arrangement: The Multifiber Arrangement, 1973 to 1980* (USITC Publication 850, January 1978), pp. 10, 11.

31 U.S., Congress, House of Representatives, Committee on Ways and Means, Subcommittee on Trade, *Background Material on the Multifiber Arrangement*, Hearing, 96th Congress, 1st Session, April 2, 1979; U.S., International Trade Commission, *The History and Current Status of the Multifiber Arrangement*; *Report on Investigation No. 332-108 Under Section 332 of the Tariff Act of 1930*, v. 1 (USITC Publication 1131, March 1981).

32 U.S., Bureau of the Census, *Annual Survey of Manufactures: Value of Product, Shipments*; *Census of Manufactures*; *Exports and Imports Related to Output; U.S. Exports: SIC Division by SIC-Based Two-Digit, Three-Digit and Four-Digit Product Code*.

33 This is the estimate of C. Fred Bergsten, Thomas Horst and Theodore H. Moran, *American Multinationals and American Interests* (Washington: Brookings Institution, 1978), p. 8). Estimates of the share of U.S. exports attributable to TNE operations vary tremendously. Besides the Bergsten, *et al.* estimate, the ITC computed a 31% figure (1970); Helleiner 46% (not dated) and Barker 66% (1966). These latter estimates, unlike the Bergsten, *et al.* figure, are most likely inflated by including all exports of firms engaged in foreign direct investment whether or not the shipments were made to their own subsidiaries. Betty L. Barker, "U.S. Foreign Trade Associated with U.S. Multinational Companies," Survey of Current Business, v. 52, n. 12 (December 1972); G.K. Helleiner, "The New Industrial Protectionism and the Developing Countries," in *Protectionism or Industrial Adjustment?*, Helleiner, et al. (Paris: Atlantic Institute for International Affairs, April 1980); U.S., Congress, Senate, *Implications of Multinational Firms for World Trade and Investment and for U.S. Trade and Labor*, by U.S. Tariff Commission, February 1973.

34 Blackhurst, *et al.*, *Trade Liberalization, Protectionism and Interdependence*, p. 63.

35 For a review of these contending arguments and effects of foreign direct investment see Bergsten, et al., *American Multinationals and American Interests*, Chapter 3.

36 For a striking advocacy and empirical analysis of this position see G.K. Helleiner, "Transnational Enterprises and the New Political Economy of U.S. Trade Policy," *Oxford Economic Papers*, v. 29, n. 1 (March 1977). See also

Blackhurst, *et al.*, *Trade Liberalization, Protectionism and Interdependence* , p. 63.

[37] *Ibid.*; Robert G. Hawkins, "U.S. Multinational Investment in Manufacturing and Domestic Economic Performance," *Occasional Paper*, n. 1 (Washington: Center for Multinational Studies, 1972).

[38] Ordinal values for all 129 sectors on the six alternate variables are available from the author.

Chapter III

U.S. Machine Tool Trade Relations

INTRODUCTION

By making comparisons across 129 U.S. industrial sectors, Chapter II demonstrated the constraining effects of export dependence on trade protectionism. These effects are now to be more concretely delineated in an illustrative case study of the U.S. machine tool industry in this and the next chapter.

The full criteria for selecting the machine tool industry for case study will be documented at a later point, but the major rationale can and should be established at the outset. Quite simply, the machine tool industry constitutes a crucial case sector as discussed in Chapter II. Like the larger industry group within which it falls, metalworking machinery (SIC 354), the machine tool industry (SIC 3541, 3542) experienced both high export dependence and high import penetration. In the late seventies, approximately 27% of U.S. machine tool sales were exported[1] — well above the national average of 7% (1979 figure).[2] As hypothesized, such relatively high levels of export dependence should have imposed severe constraints on the industry's protectionist tendencies.

Yet at the same time, imports of foreign machine tools have provided the domestic industry with solid grounds for demanding protection. Between 1975 and 1979, import penetration levels rose from approximately 18% of the U.S. market to over 29%[3] — levels that, again, far exceed the 8% level for the U.S. as a whole (1979 figure).[4] By 1982, the industry claimed that 36% of the domestic machine tool market had been captured by foreign suppliers.[5]

The machine tool case presents, then, hypothesis-testing conditions of simultaneous motives for and constraints against protectionist behavior. As will be demonstrated, the machine tool industry did not go down the protectionist road in the 1970s as did other industries — notably steel and automobiles — which confronted similar if not *lower* foreign import penetration. And a strong case can be made that dependence upon the world markets for machine tool sales served to inhibit the industry from going down that road. Before making this argument, we must first establish the theoretically interesting "problem" that cannot be completely resolved without reference to the industry's export dependence; the problem being the disjunction between the serious inroads made by foreign machine tool builders into the U.S. market and the domestic builders' relatively liberal trade policy views. Chapter IV then goes on to reconcile this disjuncture by offering export dependence as an explanation. Chapter IV also argues that as the machine took industry's export dependence changed in the 1980s so too has its trade policy views. We begin with a brief description of machine tools and the U.S. industry.

MACHINE TOOLS AND THE U.S. INDUSTRY

Machine tools are stationary power-driven machines designed to cut or shape metal. There are two general categories of machine tools: metal-cutting and metal-forming. Metal-cutting machine tools contour metal parts by cutting away unwanted portions. Basic cutting machine tools include lathes (or turning machines), boring machines, milling machines, drilling machines, and grinders.[6] Metal-forming machine tools reshape metal by hammering, punching, pressing, or bending at tremendous pressures. Important forming machine tools include forging machines, presses, shearing machines, and bending machines.

Modern machine tools make a vast array of metal parts, components, and final products.[7] Their products range from small machined pins to drilled and bored cylinder engine blocks to precision contoured airframes. The automobile industry is the largest user of machine tools in the U.S.; other major users being the aircraft, energy, construction, and farm machinery industries as well as the small machine/job shops.

Machine tools are especially valuable for making other machines — including other machine tools — thereby earning the reputation for being the "master tools of industry."[8] Indeed, almost every manufactured product is dependent upon machine tools — being made either directly by a machine tool or indirectly on machines built with the use of machine tools. As a result, "the extent of a nation's industrial development and wealth is largely determined by the machine tools it possesses."[9]

In a more narrow if not perhaps more critical sense, the machine tool industry has been considered vital to national security. As observed by General George Marshall during the Second World War:

> Practically every problem concerned with the production of arms and equipment, ships and planes starts with the question of machine tools. The tool builders, therefore, constitute the keystone of the entire procurement structure.[10]

Such views have long been held in the industry and have increasingly been adopted in government circles. As will be discussed at a later point, the link between machine tools and national security enters prominently into trade policy issues.

Despite its critical economic position, the U.S. machine tool industry — like other machine tool industries throughout the industrial world — is very small compared to other industries. The industry's $5 billion of sales in 1981 accounted for only one-quarter of one percent of all manufacturing sales in the U.S.[11] Some 80 *individual* U.S. corporations are larger than the *entire* machine tool industry.[12]

It almost naturally follows that machine tool firms, like the industry, are relatively small. According to the 1977 *Census of Manufactures*, the average machine tool establishment employed only 62 workers — a one-third drop from 93 workers in 1967.[13] Over two-thirds of all machine tool establishments had fewer than 20 employees. Only nine establishments had 1000 or more workers and only 36 employed more than 500.[14] The largest U.S. machine tool builder, Cincinnati Milacron, ranked only 322nd in the Fortune 500 list of largest corporations in 1982.

Although this small industry encompasses some 500 firms, a dozen or two companies dominate the U.S. business. In 1977 the top eight firms accounted for 47% of all machine tool shipments and the top twenty accounted for 64%. (The ten largest firms, as they stood in 1980, are listed in Table 3.1.) Still, the industry is highly competitive both among the larger firms and throughout the industry's ranks. The top four firms represented only 34% of machine tool production.[15] The U.S. machine tool industry cannot be characterized as oligopolistic.

Despite the entrance of several conglomerates into the business and the diversification outside of machine tools by some builders, the machine tool business remained largely self-contained through the 1970s. In 1977 some 90% of all machine tools were produced by firms primarily devoted to tool building, and machine tools represented 88% of the shipments of primary producers.[16]

Before moving on to the foreign trade relations of the industry, the following discussion briefly introduces four significant machine tool product innovations to which reference will often be made: numerical controls, machining centers, transfer lines, and flexible manufacturing systems.

First developed in 1949, *numerical controls* (NC) are considered to be the greatest innovation of this century in machine tool technology and a major breakthrough in manufacturing technology generally. A numerical control unit can automatically direct the operation of a machine tool by numerically coded instructions on punched tapes or computer programs. Compared to manually guided machine tools, NC tools are more accurate and consistent, and more flexible given its reprogrammability. These attributes significantly reduce machine tool downtime and offer up to ten times the productivity of manually controlled tools.[17]

Machining centers are multifunction NC machine tools that can, for instance, bore, drill, and mill a workpiece all in a single machine. Machining centers are highly versatile and can be easily reprogrammed to produce a diversity of products. *Transfer lines* are composed of an integrated series of different machine tools, sometimes up to and over 50 in number, designed to produce complex products in relatively high volume. Transfer lines, the machining equivalent of assembly lines, are very efficient and are dedicated to the production of a selected range of products.

Table 3.1

Top Ten Machine Tool Builders by 1980 Sales

Firm (Subsidiaries)		*1980 Sales($m)/Sales Category*	
1.	Cincinnati Milacron Cincinnati, Ohio	$563	machine tools
2.	Bendix Southfield, Michigan (Warner & Swasey)	623	industrial
3.	Cross & Trecker Bloomfield Hills, Michigan (Cross; Kearney & Trecker)	355	total
4.	Ex-Cell-O Troy, Michigan	387	industrial
5.	Acme-Cleveland Cleveland, Ohio (LaSalle Machine Tool; National Acme; Shalco Systems; Hillyer; Sheldon Machine)	257	capital equipment
6.	Giddings & Lewis Fond du Lac, Wisconsin	222	machine tools
7.	Textron (Bridgeport Machines; Waterbury Farrel)	250*	machine tools
8.	Ingersoll Milling Machine Rockford, Illinois	200*	total
9.	F. Jos Lamb Warren, Michigan	274	total

(continued)

Table 3.1 (continued)

Firm (Subsidiaries)	*1980 Sales($m)/Sales Category*	
10. Litton Beverly Hills, California (Landis Tool; Lucas Machine; New Britain Machine; Eldorado Tool & Manufacturing; Reed Rolled Thread Die)	190*	machine tools

* Estimate

Sources: *American Machinist*, August 1981, pp. 59-60; *Annual Reports.*

Transfer lines achieve high efficiency through relatively long production runs, but can be reprogrammed to manufacture different products only at a high cost given the large number of machines involved and the integrated nature of the system. In contrast, while machining centers can be easily reprogrammed, their production rates lag due to lack of specialization.

Flexible manufacturing systems (FMS) began to be designed in the 1970s to combine the assets and reduce the liabilities of machining centers and transfer lines. FMS are composed of several NC machine tools (either single-purpose or machining centers), a mechanism to transfer and load workpieces between and onto the machines, and an overarching computerized system to coordinate the operation of the machine tools and conveyance system. The conveyance mechanism has increasingly involved the use of industrial robots capable of workpiece handling. Because they offer adaptability to changing product requirements along with higher productivity rates, the development and production of FMS is in the forefront of machine tool research as well as industrial competition (both domestic and international).

NC machine tools along with machining centers, transfer lines, and FMS have become priority items for U.S. machine tool production. Table 3.2 shows how NC machines have come to represent increasing shares — 26% in 1980 — of total U.S. machine tool production. Machining centers and transfer lines constitute the lion's

Table 3.2

Numerically Controlled Machine Tools as a Percentage of Total Machine Tool Shipments ($ value)

1970	13.4%
1971	14.48
1972	13.40
1973	15.22
1974	17.48
1975	20.98
1976	23.00
1977	20.25
1978	20.66
1979	22.94
1980 *	26.17

* Provisional

Source: U.S., Bureau of the Census, "Current Industrial Report, Series MQ-35W, Metalworking Machinery."

share (81% in 1980) of the industry's special or "other" metal-cutting category (SIC 35417). And as presented in Table 3.3, "other" machine tools have come to dominate U. S. production. While lathes, boring, drilling, gear cutting, grinding and milling machines have diminished in importance, "other" types have risen from 17% of U.S. metal-cutting production in 1954 to nearly 40% in 1980. As will be subsequently established, NC machine tools and these other sophisticated machines have come to play prominent roles in U.S. machine tool trade as well as production.

Table 3.3

Shipments of Metal Cutting Machine Tools By Class of Product, 1954 to Date (Percentage of Dollar Value)

Year	35411 Boring Machines	35412 Drilling Machines	35413 Gear Cutting & Finishing	35414 Grinding & Polishing	35415 Lathes	35416 Milling Machines	35417 Other Metal Cutting
1954	13.0%	9.8%	5.5%	17.6%	23.5%	13.9%	16.7%
1955	11.6	8.1	5.5	18.6	23.0	13.5	19.7
1956	11.4	7.3	5.5	19.4	24.3	13.0	19.0
1957	11.6	6.4	4.4	18.2	22.7	17.6	18.9
1958	11.1	6.5	4.9	19.3	21.3	19.3	17.6
1959	9.9	7.1	5.3	20.0	25.5	15.6	16.6
1960	8.7	7.2	6.4	21.7	25.1	14.7	16.2
1961	9.7	6.9	6.9	22.8	23.8	13.9	16.1
1962	9.8	8.5	6.7	20.4	24.3	15.3	16.0
1963	10.3	7.9	6.0	20.8	23.5	16.1	15.3
1964	8.5	8.3	5.4	20.1	22.7	14.4	20.7

(continued)

Table 3.3 (continued)

Year	35411 Boring Machines	35412 Drilling Machines	35413 Gear Cutting & Finishing	35414 Grinding & Polishing	35415 Lathes	35416 Milling Machines	35417 Other Metal Cutting
1965	8.8	7.9	5.3	20.4	22.4	13.7	21.4
1966	9.0	8.6	4.8	19.1	23.3	16.4	18.8
1967	8.6	8.9	4.7	18.0	24.3	17.1	18.4
1968	7.2	5.2	5.2	18.0	24.1	17.9	22.5
1969	6.5	5.4	5.3	18.3	23.2	14.9	26.4
1970	5.7	6.5	6.0	20.0	21.9	10.0	30.0
1971	5.0	6.1	7.0	20.0	24.5	7.6	29.8
1972	6.3	7.7	5.9	17.6	23.7	9.5	29.4
1973	5.4	7.4	4.5	17.1	24.4	8.5	32.6
1974	5.3	5.1	4.3	16.4	24.3	9.0	35.6
1975	4.7	5.2	3.6	15.3	25.9	7.5	37.8
1976	4.9	6.3	3.5	15.5	27.0	8.4	34.3
1977	4.7	6.0	3.4	15.2	24.4	8.9	37.3
1978	5.0	6.5	4.4	14.9	23.0	8.3	37.9
1979	5.8	5.0	3.7	15.6	21.4	9.2	39.3

(continued)

Table 3.3 (continued)

Year	35411 Boring Machines	35412 Drilling Machines	35413 Gear Cutting & Finishing	35414 Grinding & Polishing	35415 Lathes	35416 Milling Machines	35417 Other Metal Cutting
1980(p)	5.2	5.0	3.5	15.2	22.3	9.2	39.6

p—Preliminary

Note: Individual percentages may not add to 100% due to rounding.

Source: U.S. Department of Commerce, "Current Industrial Report, Series MQ-35W, Metalworking Machinery" (Quarterly and Annual Summaries)

IMPORT PENETRATION

In order to apply the export dependence thesis to the machine tool industry, the first task is to establish the anomalous characteristics of the industry's foreign trade relations — characteristics that can only be fully understood by reference to the industry's dependence upon exports. An anomaly can be viewed as behavior generally deemed irregular given conditions that would seem to impel very different and more "normal" behavior. Later in this chapter we review what could largely be considered to be the machine tool industry's abnormal behavior on foreign trade issues; namely, the adoption of relatively liberal commercial policy positions. But first we establish the major conditions that make such behavior so anomalous: the substantial penetration of foreign machine tools into the U.S. market.

Import Trends and Suppliers

Figures 3.1, 3.2 and Table 3.4 display import penetration levels for all machine tools, for metal-cutting machine tools (which account for 71% of machine tool production, 63% of machine tool exports, and 82% of machine tool imports), and for seven major categories of cutting tools. In each case, two import penetration series are presented. The *product* series matches foreign imports against the grand total of machine tools produced (in the appropriate categories) by all U.S. manufacturers, be they primary or secondary tool builders. By calculating all imported products against all domestically produced products, this first series is the economically more accurate of the two series. In contrast, the *industry* series is computed on the basis of machine tool shipments data compiled by the industry's trade association, the National Machine Tool Builders' Association (NMTBA). This second import penetration series is widely used within the industry, is more publicized in government and the general public, and, hence, is more politically relevant than the product series. In this discussion, references to specific import penetration levels are taken from the industry series, while the product series is used for indications of import penetration trends.[18]

Figure 3.1

Import Penetration*: All Machine Tools

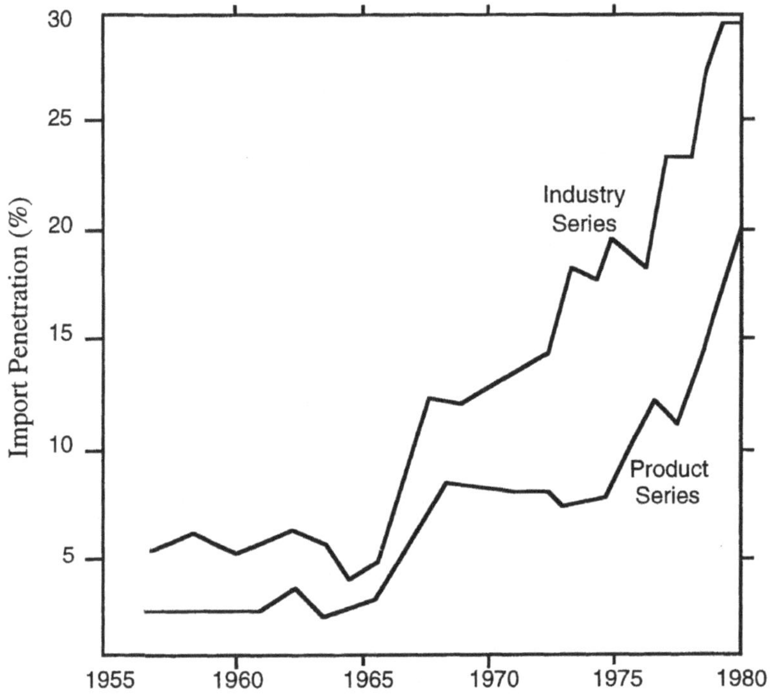

* ratio of imports to shipments plus imports minus exports

Sources: Product Series – U.S., Bureau of the Census, Annual Survey of Manufactures: Value of Product Shipments; U.S. Imports: Consumption & General, SIC-Based Products by World Areas, FT 210

Industry Series – U.S., Bureau of Economic Analysis, Business Statistics; Survey of Current Business; U.S., Bureau of the Census, U.S. Imports for Consumption: TSUSA Schedule, IM 146

Figure 3.2

Import Penetration: Metal-Cutting Machine Tools

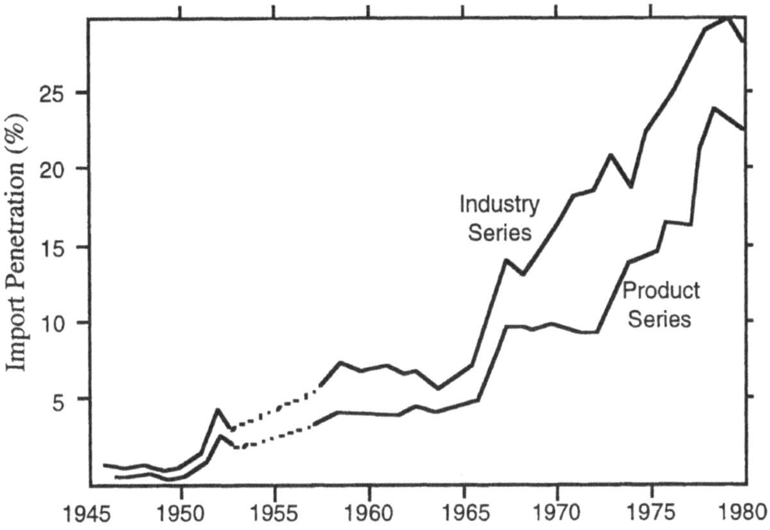

Sources: see Figure 3.1

Table 3.4

Import Penetration: All Machine Tools, Metal-Cutting Machine Tools, and Metal-Cutting Product Classes (%)

(a) Industry Series

	Metal-Cutting Product Classes								
	All Machine Tools	Metal-Cutting Machine Tools	Boring Machines	Drilling Machines	Gear-Cutting Machines	Grinding Machines	Lathes	Milling Machines	Other Machines
1946		0.40							
1947		0.32							
1948		0.43							
1949		0.38							
1950		0.65							
1951		1.46							
1952		3.66							
1953									
1954									
1955									
1956									
1957									
1958	5.97	4.95	7.73	4.18	5.07	3.48	4.82	2.71	5.34
1959	6.33	6.23	7.64	5.10	9.40	3.18	6.67	3.28	7.40
1960	6.08	5.98	8.93	4.79	8.88	3.93	7.14	4.28	6.01
1961	6.31	6.07	6.22	4.49	8.04	4.34	8.31	4.31	6.86
1962	6.76	6.51	5.20	2.00	5.71	4.01	6.70	2.31	6.08
1963	6.26	6.34	5.20	2.14	4.20	4.05	6.40	2.27	6.94
1964	5.02	5.62	6.96	2.50	4.70	3.61	5.23	2.97	4.37
1965	5.71	6.62	7.08	2.99	7.69	3.56	6.30	4.28	5.12
1966	8.87	10.39	11.47	4.86	5.26	5.93	12.24	8.28	8.12
1967	11.88	13.16	16.69	6.36	8.20	6.90	13.25	12.34	11.06
1968	11.72	12.34	19.96	14.23	8.37	7.62	11.36	12.40	7.83
1969	12.54	13.40	23.46	9.38	5.72	8.20	12.47	13.33	6.97
1970	13.23	14.83	34.39	6.54	10.61	8.21	11.68	12.85	5.61

(continued)

Table 3.4 (continued)

(a) Industry Series (continued)

	Metal-Cutting Product Classes								
	All Machine Tools	Metal-Cutting Machine Tools	Boring Machines	Drilling Machines	Gear-Cutting Machines	Grinding Machines	Lathes	Milling Machines	Other Machines
1971	14.49	15.91	32.31	7.67	15.55	9.52	10.73	17.67	7.51
1972	17.12	16.49	24.07	7.10	18.15	9.51	11.17	9.92	6.92
1973	16.67	16.63	20.31	6.71	14.42	11.17	14.26	12.04	6.35
1974	18.41	18.85	30.00	11.92	22.38	14.13	17.73	13.88	7.29
1975	17.97	17.17	39.03	10.65	38.05	13.41	15.47	17.01	7.41
1976	21.78	20.78	40.00	14.99	35.81	14.52	16.74	17.75	7.66
1977	22.08	22.53	44.50	17.47	28.85	14.40	22.43	16.73	8.32
1978	26.79	28.21	40.70	11.89	34.74	20.27	30.83	22.56	15.10
1979	29.75	30.11	32.78	15.75	32.94	16.45	37.19	24.48	15.30
1980	29.71	29.32	32.47	14.20	37.41	22.49	33.33	30.13	15.10

(b) Product Series

1946									
1947		0.16							
1948									
1949		0.21							
1950		0.28							
1951		0.93							
1952		2.57							
1953		2.06							
1954									
1955									
1956									
1957									
1958	3.20	2.95							
1959	3.22	3.42							
1960	3.22	3.37							

(continued)

Table 3.4 (continued)

(b) Product Series (continued)

	Metal-Cutting Product Classes								
	All Machine Tools	Metal-Cutting Machine Tools	Boring Machines	Drilling Machines	Gear-Cutting Machines	Grinding Machines	Lathes	Milling Machines	Other Machines
1961	3.23	3.43							
1962	3.22	3.53							
1963	3.73	4.10	5.34	2.00	3.99	3.78	6.02	2.13	4.22
1964	2.94	3.95	6.64	2.46	4.58	3.41	4.95	2.94	3.85
1965	3.85	4.70	6.90	2.92	7.47	3.38	5.97	4.11	4.20
1966	6.12	7.66	11.02	4.65	5.16	5.70	11.62	8.25	5.72
1967	7.90	9.28	15.46	6.07	7.64	6.29	12.34	11.28	7.35
1968	7.67	9.19			8.32	6.70	10.94	11.42	5.85
1969	7.46	8.81			5.35	6.66		11.58	5.44
1970	7.39	9.03	22.11	4.27	8.69	7.51	11.36	8.92	5.22
1971	7.06	8.84	18.77	3.90	11.56	7.75	11.50	11.06	6.20
1972	7.42	8.92	22.26	6.28	14.42	8.41	11.48	9.32	5.73
1973	8.33	10.14	18.95	5.99	11.90	9.68	14.56	11.22	5.44
1974	10.39	13.05	28.80	11.43	20.66	13.00	16.68	13.35	6.16
1975	11.80	13.70	36.70	10.06	33.93	13.25	15.18	16.11	6.08
1976	12.71	14.53	39.08	15.19	31.83	14.74	16.06	16.36	6.53
1977	11.64	14.34	41.44	17.21	25.71	13.65	21.59	16.00	7.77
1978	15.37	18.72	37.19	11.69	28.65	19.64	30.03	21.50	13.31
1979	18.21	21.57	33.96	14.26	31.84	17.61	36.22	26.05	14.38
1980	20.28	21.01							

Sources: see Figure 3.1

The trends are clearly ominous for the U.S. machine tool industry. Three major import surges are evident: 1950-1952, 1965-1967, and since 1973. The first surge was largely attributable to European machine tool builders who in the early 1950s established major marketing outlets in the U.S. Machine tool imports rose from $2 million in 1950 to $48 million in 1952.[19] These imports consisted of such general purpose machines as radial drills, turret lathes, engine lathes, and shapers. Metal-cutting import penetration levels rose from one-third of one percent of the U.S. market in 1949 to nearly 4% in 1952.

By 1962 U.S. builders became very concerned over an imminent onslaught of imports as rapidly expanding European machine tool capacity began to outstrip European demand.[20] The expected surge was realized when U.S. imports rose from $46 million in 1964 to $204 million in 1967.[21] Between 1964 and 1967, imports from West Germany rose 302% to lead all foreign suppliers with a 28% share of U.S. imports (1967). Over this period, imports from Switzerland jumped 115%, Britain 439%, Canada 467%, Italy 962%, and Japan 1210%.[22] The phenomenal success of Japanese suppliers was based on the marketing of low priced conventional machines through the Japan Machine Tool Trade Association and its Chicago trade center, both established in 1962.[23]

By 1967, nearly 12% of the U.S. market was captured by foreign machine tool builders. Nearly 17% of the boring machine market was penetrated by imports, and lathes and milling machines were hard hit as well. The U.S. trade balance in each of these three machine tool categories went into deficit in 1966 and have since deteriorated. In 1967, metal-cutting machine tools suffered their first negative trade balance in history.

By far the most substantial import surge has taken place since 1973 and especially since 1977. U.S. machine tool imports have risen from $167 million in 1973 to $401 million in 1977 to $1483 million in 1981.[24] U.S. trade in drilling machines went into deficit in 1976, grinding machines in 1979, and machining centers in 1980.[25] For the first time in industry history, trade in all machine tools, both cutting and forming, went into deficit in 1978 and the deficit has since worsened. The share of the U.S. market captured by foreign suppliers jumped from 17% in 1973 to 29% in 1980. 1980 import penetration levels for boring machines stood at 32%, for gear cutting at 37%, and for lathes at 33%. The import share for the all-important NC tool market stood at 25%, and foreign penetration for machining centers has risen from 6% in 1973 to 21% in 1980. In unit terms, the industry claimed a 36% import penetration level for 1981.

Established European suppliers have bolstered their positions in the U.S. market by moving into more sophisticated product lines while Indian, Spanish, Rumanian and especially Taiwanese and South Korean builders have moved into the market for general purpose machines. By the 1980s, Taiwan and South Korea began to successfully market low-cost, light-duty NC machine tools in the United States.[26]

But far and away the most significant foreign supplier of machine tools has been Japan. U.S. imports from Japan have skyrocketed from $22 million in 1973 to $106 million in 1977 to $688 million in 1981. In 1977, Japan overtook West Germany as America's number one foreign supplier, and by 1981 Japanese machines accounted for over 46% of all U.S. machine tool imports.[27]

Japanese inroads into the U.S. market have been fairly broad based, encompassing many product classes. Yet spearheading the influx has been series-produced, price-competitive NC lathes and NC machining centers. These two product classes accounted for 70% of all Japanese metal-cutting shipments to the U.S. in 1980.[28] Japanese builders have come to capture 32% of the U.S. market for NC lathes and 27% for machining centers; corresponding import penetration levels in unit terms are 59% and 50%, respectively (all 1980 figures).[29] Altogether nearly three quarters of U.S. imports from Japan consist of NC machines.[30]

As stated by the NMTBA, "Clearly, the Japanese have targeted the United States machine tool market."[31] In 1975, just under 23% of all Japanese export shipments were destined for the U.S. while by 1980 that figure reached nearly 40%.[32] By 1980, Japan claimed 10% of the entire U.S. market on its own; West Germany, Britain and Taiwan — the next three major foreign suppliers — together captured another 9% of the market.[33]

In conclusion, foreign machine tool builders came to capture one-third of the U.S. market and increasingly penetrated the higher, more sophisticated end of the market. This latter point will be elaborated in a later section, but before doing so it is important to briefly discuss the reasons for the success of foreign machine tools in the U.S. market.

Reasons for Import Penetration

Successful import penetration can be generally viewed as being based on supply-side or export push factors and/or on demand-side or import pull factors.[34] Export push factors concern actions on the part of foreign suppliers to enhance their shipments abroad. Subsidization of export industries, preferential export financing, aggressive marketing tactics, pricing practices, and the export promotion activities of governments and trading companies are among the methods by which foreign countries can "push" their exports. The prior section mentioned some of the export push tactics that have played a part in the penetration of the U.S. machine tool market; namely, the establishment of marketing outlets in the U.S. by European builders in the early 1950s, and the creation of the Japan Machine Tool Trade Association and its Chicago trade center in 1962. In contrast, this section will focus on the import pull factors operating in the U.S. machine tool market.

Import pull factors generally concern the nexus between strong domestic market demand and domestic suppliers falling short of meeting demand. When consumption needs exceed domestic supply, quantitatively or qualitatively, foreign imports may be sought to fill the gap. In this context, the attributes of foreign products and not their marketing become decisive in whether they are "pulled" into another market.

Three major product characteristics have facilitated the import pull of foreign machine tools into the U.S. market: delivery, price, and quality. Even if all export push tactics were eliminated from machine tool trade, the prevalence of these import pull factors would certainly sustain the import trends discussed in the prior section.

Delivery. Traditionally foreign suppliers have made their inroads into the U.S. market when domestic machine tool demand booms. In virtually every high growth period, domestic demand for machine tools has outstripped the abilities of U.S. builders to expand productive capacity. Backlogs of unfilled orders build up with domestic firms and customers begin to look overseas to meet their machine tool needs. This phenomenon has occurred time and time again.

Mobilization for the Korean war quickly strained domestic machine tool capacity by mid-1951. In response to this bottleneck, machine tool users precipitated the first major import surge of the early fifties by making their purchases with West German and other foreign suppliers.

During the next machine tool boom period of the mid-sixties, backlogs piled up to the extent that buyers had to wait up to 18-24 months for delivery.[35] Such backlogs opened the door for foreign machine tool builders offering quicker delivery time giving rise to the second import surge.

The third import surge was also fueled by the growth of domestic backlogs. Through the late seventies, U.S. builders could again quote only 18-24 month delivery dates.[36] In contrast, foreign builders were able to promise delivery in a single week to three months.[37] Japanese producers have been particularly acclaimed for their efficient inventory control and quick delivery. Unlike U.S. machine tool builders, Japanese builders do not await orders to begin production. The Japanese plan in advance on certain levels of output, earmark a percentage for export, and then build their machines in greater volume than U.S. producers. They then stockpile their machines in U.S. warehouses and are able to deliver on demand or within two to three months.[38] Figure 3.3 illustrates what has been a close correlation between levels of imports and unfilled orders.

This correlation held up until the mid-1970s. Just as imports rise when backlogs mount, imports have usually suffered when domestic deliveries become more forthcoming as machine tool business slackens. The latter was the case in the mid-fifties and the late sixties-early seventies. By the mid-1970s, however, this relationship began to break down. When 1975 domestic backlogs fell nearly 60% (in constant dollar terms) from 1974 levels, foreign import penetration, instead of diminishing, more than held its own. The most significant break took place with the machine tool recession that began in mid-1980. From a backlog high of 20 months at the end of 1979, unfilled orders amounted to only six months' work in March 1982.[39] Despite increasing delivery-competitiveness of domestic builders, import levels showed little sign of waning. While most foreign builders still had delivery advantages over domestic builders, other factors — such as price and/or quality — underlay this success of foreign imports.

Figure 3.3

Machine Tool Industry Indices of Unfilled Orders and Imports
Cutting and Forming 1967 = 100(a) Ratio Scale

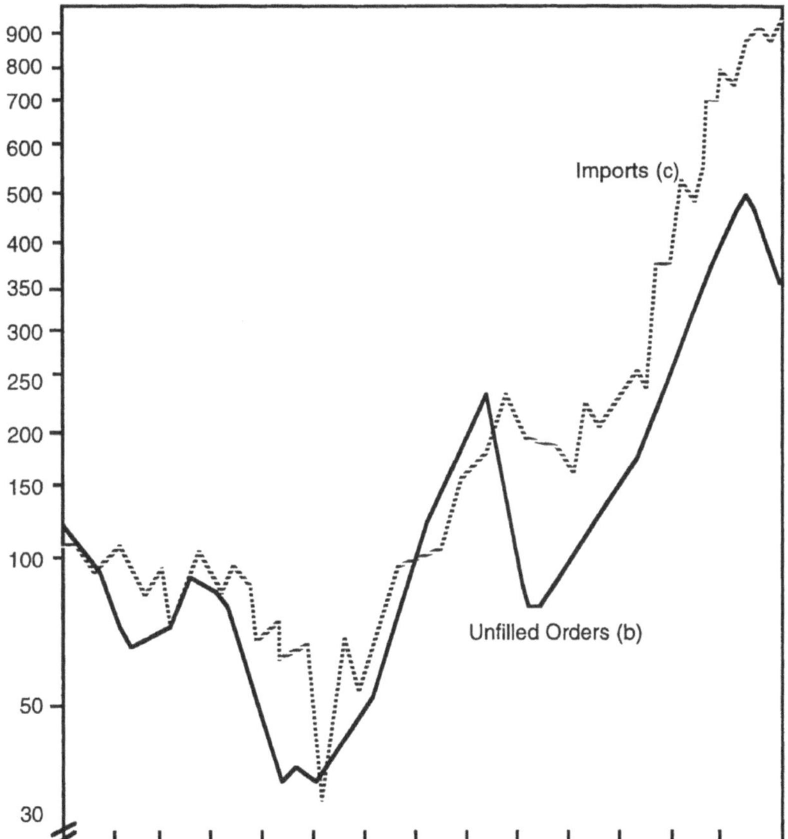

(a) 1967 quarterly average
(b) Unfilled orders series based on end-of-quarter values
(c) Imports quarterly

Source: U.S. Bureau of the Census "FT 135" and National Machine Tool Builders Association

Price. Beginning as early as the 1920s, U.S. machine tool builders complained about foreign machine tool prices being 20-50% below their own.[40] Through the 1950s, the culprits were European and especially West German machines. Since the mid-1960s, they have been Japanese tools and later Taiwanese and South Korean. When combined with quality equal to that of U.S. tools or at least sufficient to "get the job done," price advantages have made foreign machine tools irresistible to even the largest, most discriminating U.S. buyers.[41]

In 1980, average prices for imported small-to-medium NC lathes were $60,000 - $82,000 compared to $109,000 - $219,000 for U.S. equivalents. Average price tags for small and medium-sized machining centers stood at $82,000 - $145,000 for imports versus $139,000 - $313,000 for U.S.-built machines.[42] The Japanese have been particularly credited for their cost-efficient, relatively high-volume production runs for even NC machine tools. In contrast, U.S. builders traditionally built machines only on a customized, small (if not single) order basis that prevents taking advantage of unit price reduction through larger production runs.

While many foreign machine tool builders, particularly in developing countries, maintained a price edge over U.S. builders, such advantages drastically diminished if not disappeared altogether for America's industrialized competitors. Beginning in the mid-1970s, major European machine tool builders — West German, British, Swedish, Swiss and Italian — saw any price edges blunted.[43] From 1977, prices on Japanese tools also rose significantly. Due in large part to the 1977-78 depreciation of the dollar against the yen, major machine tool exporters to the U.S. were forced to increase prices 8% at the lower end of their product lines and as much as 30-35% at the high end.[44]

On the question of Japanese pricing, a major U.S. distributor of Japanese tools observed:

> Long gone are the days when Japanese-built machinery was bought on price alone. The machines we're selling now are being sold on the basis of superior workmanship and superior performance.[45]

These views were confirmed in a major study of Japanese machine tool exports to the U.S. by Wilbur F. Monroe:

> To date there has been no appearance of strong evidence indicating that comparatively lower prices

> of Japanese machine tools have provided an important advantage. A significant price differential between comparable U.S. and Japanese machine tools apparently existed until rather recently. But price tends to be much less important in the determination of a sale than product quality and design, dependability and availability of parts and service.[46]

Quality. As observed in the above quotations, product and service quality increasingly fueled the penetration of imports into the U.S. — just as both delivery and price advantages (while not non-existent) diminished in importance. While foreign machine tool builders in developing countries had yet to establish reliable quality and overseas servicing,[47] European and the all-important Japanese machine tools achieved high market acceptance in the U.S..

In a 1980 survey by one of the industry's major trade publications, Japanese machine tools were rated at least as reliable and serviceable as U.S. tools.[48] Japanese as well as European machine tool builders fared even better in an independent survey, also conducted in 1980, by Frost and Sullivan, the New York consulting firm. Imported machines were rated superior to domestic counterparts in 52% of designated machine tool categories, inferior in 32%, and equal to domestic tools in 16%.[49]

Regarding service, the Japanese have particularly excelled. Utilizing U.S. nationals, Japanese builders developed extensive service and parts networks across the country. Such facilities have been critical to Japan's success in exporting NC machine tools since such machines require significant after-sale service.[50] These favorable attributes of foreign machine tools have been all the more valued given not insignificant complaints about U.S. machine tool quality, reliability, and servicing — complaints that often characterize Japanese machine tool builders favorably.[51]

The astonishing market success of Japanese NC lathes and machining centers was based on all three factors discussed above and more. U.S. and European machine tool builders have long been leaders in larger, heavy duty machines. While suppliers, domestic and foreign, for conventional, lighter-duty tools have never been lacking, U.S. builders — until the 1980s — offered little in the way of mid-range product. Japanese suppliers exploited this market gap with their push for cost efficient, medium-sized, highly versatile NC lathes and machining centers. Japanese builders excelled in this intermediate product range and efficiently served what was a latent U.S. market with standardized high-production runs.[52]

In sum, price and delivery advantages have long aided the success of foreign machine tools in the U.S. market. Yet the bulk of more recent imports — particularly those from the number one foreign supplier, Japan — were not purchased simply because they were cheap and happened to be readily available in warehouses. Instead, the quality of the foreign product and of the service that goes with it increasingly became responsible for their market acceptance. U.S. machine tool builders tried to recoup on the price and delivery fronts. Yet the quality-based success of recent foreign machine tools has made the high import penetration levels all the more formidable and less vulnerable. One purpose of the following section on sophisticated machine tool imports is to augment the quality theme with specific examples.

Penetration at the High End of the Market

In late 1977, James A. Gray, President of the NMTBA, sounded a clear warning:

> In the past our foreign competitors concentrated on the U.S. market for standard, general-purpose machine tools. They did not try to compete with our numerical controlled, computer controlled (CNC), and other more sophisticated machine tools. During the past year or so the reports and complaints . . . have changed. The foreigners have recently drawn ahead on the low-cost end of the NC and CNC market; and I believe that this new effort is only a first step in a well-designed program aimed at our domestic market for sophisticated machine tools.[53]

By 1977, the writing was clearly on the walls and Mr. Gray's fears have since been thoroughly realized. This section will more fully specify the market penetration by sophisticated foreign machine tools. In doing so, flesh will be given to the statistics presented above and illustrations provided for the sources of the import successes discussed thus far. Import penetration in two specific technologies and three sophisticated markets will be discussed: NC machine tools, the automobile industry, the aerospace industry, defense procurement, and flexible manufacturing systems.

NC Machine Tools. The U.S. market for numerically controlled tools became no less penetrated than the general machine tool market. By 1980, 25% of the market for all NC tools, 29% of the NC lathe market, and 21% of the machining center market were lost to foreign suppliers. In unit terms, these import penetration levels stood at 41%, 51% and 34%, respectively.[54] In the first nine months of 1981, foreign builders captured 44% of all machining center sales in value and 51% in units.[55] Major Japanese suppliers of NC lathes included Yamazaki Machinery Works, Mori Seiki, Okuma Machinery Works, Ikegai Iron Works, Hitachi Seiki, Wasino Machine, and Takisawa. Makino Milling Machine, Toshiba Machine and Toyada Machine Works of Japan and Maho Werkzeugmaschienenban and Wotan of West Germany achieved considerable success in marketing machining centers in the U.S.

A foreign supplier even began to equip *U.S.* machine tools with numerical controls. General Numeric, a U.S.-based joint venture between Fujitsu Fanuc of Japan and Siemens of West Germany, was the only major NC manufacturer, domestic or foreign, to increase its market share in the equipping of U.S. tools between 1980 and 1981 (from a 6% market share to 8%). Such U.S. machine tool builders as Pratt & Whitney Machine Tool (West Hartford, Connecticut) and United States Machine Tool (Hartford, Connecticut) reported that their customers were pressing for General Numeric controls because of their quality. General Numeric came to rank as the fifth largest NC builder for U.S. tools behind Allen Bradley (with a 1981 market share of 16%), General Electric (13%), Cincinnati Milacron (12%), and Bridgeport Machines (10%).[56]

Automobile Industry. Until the late seventies, U.S. machine tool builders had the domestic market for large, sophisticated machining systems all to themselves. Then in early 1979 a foreign builder dramatically broke into one of the two major markets for large systems, the automobile industry. In an unprecedented move, Ford ordered a large transfer line from Toyama Machine Works to machine lock-up torque converter housings for automatic transmissions. The $4.6 - $4.7 million order was some $500,000 less than comparable U.S.-built systems. Following the Toyama sale, European as well as Japanese builders moved into the auto industry market in big way. Ford made a $10 million order for six large stamping presses from Komatsu (Japan) following the delivery of twelve Komatsu presses to the General Motors Fisher Body Division.[57] General Motors also made milling machine purchases from Gebr. Heller (West Germany).[58] General Motors and Chrysler operated transfer lines from Comau Industriale (Italy), a subsidiary of Fiat.[59] Cummings Engine Company awarded a $15 - $20 million contract to Nabenfabrik Alfing Kessler (West Germany) for a large machining system to produce connecting rods for new small-to-medium sized diesel engines.[60] At the very least, traditional domestic machine tool suppliers for the auto industry — e.g., F. Joseph Lamb, Cross, Bendix and Ex-Cell-O — took heed that the auto manufacturers were asking foreign builders for quotes — quotes that often set new bounds for competitiveness.

Aerospace Industry. Foreign builders have even made serious inroads, again largely since 1979, into the other major U.S. market for large machining systems, the aircraft industry. A brief discussion of some of the most notable sales follows.

In 1979 Yamazaki Machinery Works gained orders for 14 horizontal machining centers from General Electric for its jet engine plant. Yamazaki won the contract over domestic builders because it had a nearby assembly plant, was willing to modify its machines to meet General Electric's specifications, use GE controls, and still make quick delivery.[61]

Trumpf (West Germany) drilling machines were successfully marketed to McDonnell Douglas, Fairchild Republic, and Northrop. In 1981 Trumpf machines were chosen for automated sheet metal manufacturing facilities at Lockheed and Grumman Aerospace.[62] These latter imports were all the more significant since the automated sheet systems constituted explicitly designed first steps, by the aerospace firms and the U.S. Air Force, towards the fully automated, computerized "factory of the future."

Perhaps the single most effective foreign builder in penetrating the aerospace market was the Japanese firm Shin Nippon Koki (SNK). In 1979, SNK, the world's second largest producer of multispindle, multiaxis profilers (precision millers) behind Cincinnati Milacron, began making sales to Lockheed Aircraft, General Dynamics, Grumman, Avco Lycoming Division, Fairchild Aircraft, and Northrop. SNK profilers were not only 10% cheaper than and had half the lead time of Milacron machines, but they proved superior in quality and performance. Milacron profilers are end-products of incremental improvements in a basic design developed in the late sixties. In contrast, SNK machines were designed in the late seventies with one third fewer parts and almost twice the horsepower of the Milacron product.[63]

Two episodes illustrate the success of SNK. In early 1979, General Dynamics, in the midst of a $125 million modernization program to build the F-16 fighter, needed to double or triple its capacity to produce aluminum wing spars in order to meet worldwide demand. In particular, General Dynamics needed three profilers with hi-tech characteristics: three spindles, four axis, and high-speed cutting heads. The problem grew to crisis-proportions when U.S. machine tool builders quoted two year deliveries. Then in mid-1979, SNK promised not only one year delivery but, in an unprecedented move, agreed to pay penalties for late delivery. On top of this, SNK's price quotation was 20% below that of domestic builders — $2.4 million versus $3 million for the entire package. SNK not only delivered the sophisticated profilers on time, but even sent a team to Westinghouse, General Dynamics' NC subcontractor, to interface its machines with U.S. controls.[64]

In early 1980, Northrop began converting an assembly plant into a machining shop to fabricate the Navy's F-18 fighter. Again, sophisticated profilers were needed and, again, two U.S. machine tool builders quoted two year lead times. According to Northrop's fabrication manager, a two year delivery date would have jeopardized the manufacturer's ability to meet its commitments to McDonnell Douglas (the major contractor for the F-18) and the Navy. So Northrop turned to SNK and not only received a one year delivery promise but a lower price offer. To seal the deal, the Japanese firm even agreed to establish a West coast service and parts facility.[65]

Military Procurement. Since aerospace firms are major defense contractors, we have already seen that foreign machine tools have increasingly been used for military procurement. One of the most controversial military purchases involved the XM-l tank. In early 1978, Chrysler, the major contractor for the XM-l, awarded a $2.7 million subcontract to Mitsubishi Heavy Industries (Japan) for three NC horizontal machining centers with automatic tool changers to machine the tank's turret. According to Chrysler's Sterling Defense Division manager, Mitsubishi was chosen because its machines were ranked first technically and because the tools carried the lowest price. And the Mitsubishi price offer had to be exceptionally low in order to satisfy the Buy America Act requirement that foreign bids on military orders be at least 50% below the lowest U.S. bid. In the XM-l tank case, Carlton Machine Tool (Cincinnati, Ohio) made the lowest domestic offer of $6.3 million; other losing bidders were Cincinnati Milacron and Kearney & Trecker.[66]

Yet a classic case study illustrating the attributes of a foreign machine tool builder and the shortcomings of U.S. builders involved the Army's and General Electric's efforts to find sophisticated milling machines for their joint development of a high performance turboshaft engine for military helicopters.[67] Four-spindle, five-axis, high-speed milling machines were required to precision machine integrated blades and disks in the engine — an engine selected for three major military helicopter programs. In 1976, the Army chose New England Machine and Tool (Berlin, Connecticut) to develop the millers, and in 1977 General Electric purchased a similar machine from Rigid, Ltd., a relatively small and unheralded Swiss builder. While the Rigid machine performed quite well, the New England miller could not handle the high-strength, superalloy materials, the high spindle speeds, nor the long, complex machining runs.

While quite aware "that the Rigid machine could do the job quite nicely," the Army was "very anxious to develop a domestic source." "We much preferred to see this capability developed in the United States, to see a U.S. builder take up the project," recalled an Army official. Accordingly in late 1977, the Army solicited proposals from Cincinnati Milacron, White-Sundstrand Machine Tool (Belvidere, Illinois), Ex-Cell-O, Kearney & Trecker, Producto Machine (Bridgeport, Connecticut), and Cross. Whether because of overcommitments, two year lead times, or outright refusals to take the risk of work with the government, the effective response from each firm was negative. Faced with mounting pressure to get the project moving, the Army and General Electric reluctantly placed orders for six Rigid machines — even though they were more expensive than the New England millers.

Through 1978 the Rigid millers proved exceptionally productive — the extremely high finishes they were achieving were making the engines more efficient and less costly than expected. Spurred by the success of the Swiss machines, the Army and General Electric went back to the U.S. machine tool industry for quotes and, again, the response was negative. Discouraged but undaunted, the Army made a third attempt to develop its domestic source. Apparently the third time was a charm and Ex-Cell-O submitted a suitable offer. The Army has subsequently ordered machines from both Rigid and Ex-Cell-O. The military got its domestic source, but "it was like pulling teeth." Because of the U.S. industry's reluctance to innovate and assume some risk, a small Swiss company with a higher-priced product was able to become a major supplier for, as one Army manufacturing official put it, "quite simply the most modern machining system of its type in the world."

Flexible Manufacturing Systems. Foreign machine tool builders have even broken into the U.S. market for the hottest machine tool technology, flexible manufacturing systems. In fact, the *first* FMS sale to a major U.S. automaker was made by a foreign builder, Comau Industriale. In July 1980, the Italian firm sold three machining centers linked together by automated parts handling equipment to General Motors to machine differential carriers on light trucks. General Motors officials privately conveyed that the order went to Comau because U.S. builders were unresponsive to their needs and were not competitive in price or delivery.[68] In 1981 Comau was asked to bid on an FMS at the U.S. arsenal at Watervliet, New York. The arsenal constitutes the Army's focal point for the development and acquisition of FMS.[69]

West German firms also established strong positions in the U.S. FMS market. Two of Deere and Company's first three FMS were West German. One system, made by Burkhardt and Weber, was able to machine seven different tractor parts. The other FMS was a 25-machine Liebherr-Verzahntechnik gear-cutting system. (Kearney & Trecker supplied Deere's third FMS).[70] West Germany's Scharmann built an FMS for Caterpillar Tractor to machine engine roller frames at its Belgian plant.[71]

Yamazaki Machinery Works led Japan's FMS charge into the U.S. market with at least three major sales as of September 1981 of the highly innovative Yamazaki Machining System. Instead of moving workpieces among various machines, the Yamazaki system keeps the workpiece stationary and multiple, mobile spindles and tool heads are brought to bear on it.[72]

Import Penetration and Hegemonic Decline

We have seen how foreign machine tools came to capture 29% of the U.S. market in terms of value and 36% in unit terms. Although it is very difficult to decipher an impact independent of the business cycle, import penetration led to the dropping of product lines (e.g., engine lathes, shapers, and planers) and to layoffs and plant closings.[73] In keeping with the theoretical literature on import penetration discussed in Chapter I, the U.S. machine tool industry should be a classic case study of protectionism.

Indeed, the machine tool industry's import penetration levels compare closely to those of other industries which have sought and received trade protection. Table 3.5 presents foreign penetration levels for major protectionist industries along with machine tools. It can be seen that machine tool penetration ratios *surpass* those of the automobile, footwear, and television industries when the latter made their protectionist demands. And the import threat that machine tools faced *far exceeds* the threat faced by the mushroom, specialty steel, carbon steel, textile, and meat industries when they made their initial petitions for import relief.[74] Only import penetration levels for CB radios, clothespins, and bolts, nuts, and screws surpass machine tool levels. Given conventional wisdom, the machine tool industry in the 1970s seemed to be well-positioned if not *long overdue* for seeking trade protection.

Table 3.5

Import Penetration Levels for Major Protectionist Industries and Machine Tools

		Import Penetration (%)		
Industry	Date	Quan*	Value	Comments§
Citizens Band Radios	1977	91		Relief petitioned 8/77. Tariffs raised 1978-1980.
Clothespins	1977	40		ITC initiates investigation 7/78. Three year global quota.
Bolts, Nuts, Screws	1975 1981	38 25-68	21 15-46	Relief petitioned 5/75. Tariffs raised for three years.
Machine Tools	1980 1981	36	29	
Passenger Automobiles	1980	35	30	Relief petitioned 6/80. Voluntary Export Restraint, 1981-1983.
Nonrubber Footwear	1970 1980	32 49		President initiates investigation late 1970. Orderly marketing arrangements, 1977-1981.

(continued)

Table 3.5 (continued)

		Import Penetration (%)		
Industry	Date	Quan*	Value	Comments§
Color Televisions	1976 1979	33 14	20 11	Relief petitioned 9/76. Orderly marketing arrangements, 1977-.
Canned Mushrooms	1964/65 1979/80	19 53		Relief petitioned 1964. Voluntary export restraints, 1976-1979. Tariff increase 1980-.
Specialty Steel	1975 1978	18 12		Relief petitioned 7/75. Orderly marketing arrangement, 1976-.
Carbon Steel Mill Products	1971 1977 1980	18 18 16		Voluntary export restraints, 1969-1974. Trigger Price Mechanism, 1978-1982. Antidumping & Countervailing duty relief petitioned 1/82.
Textiles	1961 1979	5 11		Short-Term Arrangement negotiated 7/61. Succeeded by Long-Term Arrangement and Multi-Fiber Arrangement, 1962-.
Meat	1964 1976	1 2		Meat Import Act enacted 1964; restraints since.

(continued)

Table 3.5 (continued)

* Quantity
§ Unless otherwise indicated, relief petitioned and investigations initiated under Escape Clause provisions.

Sources: U.S., International Trade Commission, *Bolts, Nuts, and Large Screws of Iron or Steel*, November 1981, #1193; *Canned Mushrooms*, March 1976, #761; *Canned Mushrooms*, August 1980, #1089; *Certain Motor Vehicles and Certain Chassis & Bodies Thereof*, December 1980, #1110; *Certain Steel Products from Belgium, Brazil, France, Italy, Luxembourg, the Netherlands, Romania, the United Kingdom, and West Germany*, February 1982, #1221; *Citizens Band (CB) Radio Transceivers*, February 1978, #852; *Clothespins*, December 1978, #933; *Color Television Receivers and Subassemblies Thereof*, May 1980, #1068; *Live Cattle and Certain Edible Products of Cattle*, September 1977, #834; *Multifiber Arrangement, 1973 to 1980*, March 1981, #1131; *Nonrubber Footwear*, January 1971, #359; *Stainless Steel and Alloy Steel*, April 1979, #968; U.S., Office of the United States Trade Representative, *Trade Action Monitoring System*.

Besides steep import penetration levels, we have also found that inroads were increasingly being made by high quality, sophisticated foreign machine tools. Penetration by such machines meant that even the largest, most sophisticated U.S. machine tool builders were not been spared serious injury. In just the few episodes briefly discussed, we have seen how foreign builders have taken orders away from Cross, F. Jos. Lamb, Bendix, and Ex-Cell-O for transfer lines; from Cincinnati Milacron for profilers; from Carlton, Milacron, and Kearney & Trecker for the XM-l tank contract; and from Kearney & Trecker, Milacron, and Bendix for FMS. The onslaught of Japanese NC lathes and machining centers has shaken the industry as a whole and the larger builders in particular. Table 3.6 lists the U.S. builders producing these two machine tool lines. Nine of the top ten builders were among those facing and suffering from direct Japanese competition.

Table 3.6

U.S. Builders of NC Lathes and Machining Centers

	<u>*NC Lathes*</u>
	American Tool
	Bullard Company
1	Cincinnati Milacron
	Clausing Machine Tools
3*	Cross Company
	Detroit Broad & Machine
	Babcock & Wilcox
	Dunham Tool
	Farrel Machinery Group
	USM Corporation
6*	Giddings & Lewis Machine Tool
	Div. of Giddings & Lewis
	HES Machine Tool
	Hardinge Brothers
8	Ingersoll Milling Machine
	LeBlond Machine Tool
	Lehmann, Div. of Smith International

(continued)

Table 3.6 (continued)

	NC Lathes (continued)
	Lodge & Shipley
	Monarch Machine Tool
	Motch Mfg. Division Motch & Merryweather Machinery
10*	New Britain Machine Division Litton
	Pratt & Whitney Machine Tool Div. Colt
	R D & D Corporation
	Seneca Fall Machine
	Sheldon Machine
	South Bend Lathe
	Turchan
	United States Machine Tools
	Universal-Automatic
	Wadell Equipment
2*	Warner & Swasey Turning Machine Division
	Weldon Hydraulics
	Machining Centers
	Bayer Industries
	Boston Digital
7*	Bridgeport Machines Division Textron
	Carlton Machine Tool
1	Cincinnati Milacron
	DeVlieg Machine
6*	Giddings & Lewis MachineTool Div. of Giddings & Lewis
6*	Giddings & Lewis-Bickford Machine Div. of Giddings & Lewis
	Hamill

(continued)

Table 3.6 (continued)

<u>*Machining Centers*</u> *(continued)*

5*	Hillyer
	Div. of Acme-Cleveland
	Hurco Manufacturing
8	Ingersoll Milling Machine
3*	Kearney & Trecker
	Kysor Machine Tool
10*	Lucas Machine Division
	Litton
	Monarch Machine Tool
	Moog, Hydra-Point Division
	Morey Machinery
	Pratt & Whitney Machine Tool
	Div. Colt
	Sheldon Machine
	Turchan
	White-Sundstrand Machine Tool
	White Consolidated
4*	Workcenter Operation
	Ex-Cell-O

Rank in Top Ten as indicated.
* Rank of parent company

Source: NMTBA Directory 1982.

Foreign penetration on the basis of quality and technological sophistication reflects a broader, more fundamental shift in the relative international position of the U.S. machine tool industry. The loss of domestic market share to foreign builders mirrored the erosion of U.S. dominance in the world machine tool industry generally. That erosion is conveyed in world production and world trade trends.

Through 1957, four out of every ten machine tools produced in the world were built in the U.S.[75] In 1967, 34% of the world's machine tools were still made in the U.S. (see Figure 3.4). From 1967, America's relative position in the world machine tool industry has deteriorated significantly. By 1980, the U.S. share of world production had dropped to only 18%. Western European builders came to account for 38% of world production, Eastern European manufacturers for 22%, and Japanese firms for 14%.

Figure 3.4

Share of World Output
(Machine Tool Production as a Percent of World Production)

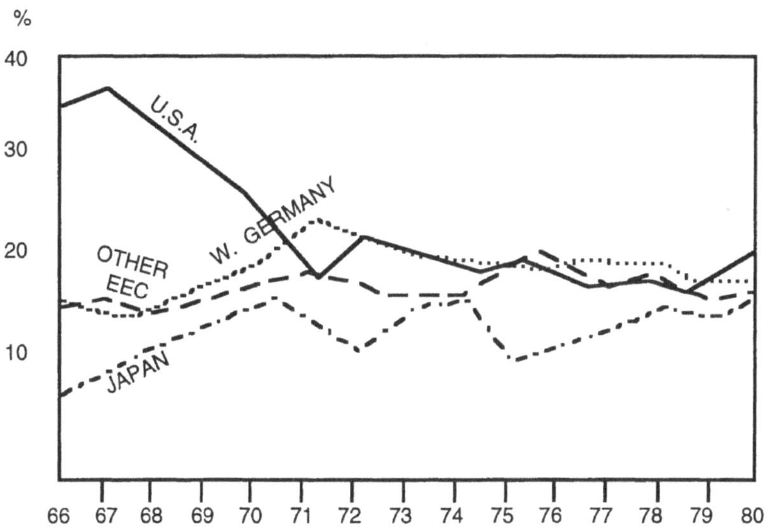

Source: American Machinist

While the U.S. never held a leading position in world machine tool trade, U.S. machine tool builders also suffered a secular decline in their share of world exports. From a 25% share of world machine tool exports in the 1940s, the U.S. share dropped to 21% in 1964 and to 10% in 1971.[76] By 1980, U.S. exports represented only 7% of world exports compared to a Western European share of 57%, an Eastern European share of 17%, and a Japanese share of 13% (see Figure 3.5).

Figure 3.5

Share of World Exports
(Machine Tool Exports as a Percent of Total World Exports)

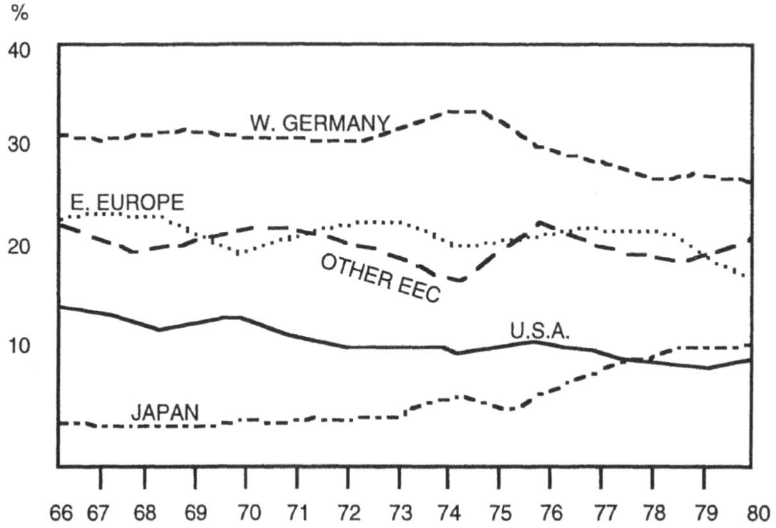

Source: American Machinist, U.S. Dept. of Commerce, NMTBA

The relative decline of the U.S. machine tool industry in terms of world production and world trade would not bode well for the industry's trade policy views according to the literature on hegemonic stability discussed in Chapter I. Hegemony stability theory, which was originally cast at the level of entire economies but has since been applied to specific issue areas and industrial sectors,[77] links the decline of economic power to the outbreak of protectionism. Given the U.S. machine tool industry's fall from grace, this literature, like that on import penetration, would peg U.S. machine tool builders as ripe for if not rife with trade protectionism. To test these expectations, we now move to an analysis of the trade policy views of the U.S. industry.

TRADE POLICY

We have thus far drawn a fairly gloomy picture of the U.S. machine tool industry. Import penetration and hegemonic decline set the primary backdrop against which we can establish the anomalous characteristics of the industry's trade policy views. Examining that anomalous behavior is the task in this section. We begin with a brief discussion of the politics of the machine tool industry, and then move on to a largely chronological analysis of the industry's positions on commercial policy issues.

Politics and Policymaking in the Machine Tool Industry

This review of machine tool trade policy is largely based upon the public statements, primarily made in congressional hearings, by staff members of and corporate executives speaking for the industry's national trade association, the National Machine Tool Builders' Association (NMTBA). The association is comprised of some 400 firms which represent nearly 90% of U.S. machine tool production.[78] Firms must produce tools within the United States for a full year to be eligible for NMTBA membership; enterprises engaging solely in the distribution and/or importation of machine tools are ineligible.

The association is governed by a fourteen-member Board of Directors elected by the general membership. Much of the association's work is guided by committees composed of member companies and appointed by the Board of Directors. One such committee, the Government Relations Committee, is solely devoted to the formulation of association positions on public policy issues and the setting of legislative/lobbying priorities — subject to approval of the Board of Directors.

The NMTBA serves as the focal point if not sole agent of U.S. machine tool politics. It has long had a full-time lobbying staff in Washington, D.C. Over the past 30 years, the association has taken frequent and vocal positions on issues dealing with labor law, taxes, depreciation, fiscal policy, government procurement as well as foreign trade. A few of the larger machine tool builders have their own lobbying staffs, but their efforts are almost always coordinated with those of the NMTBA.

Even if the association serves as the primary vehicle for machine tool politics, the question arises as to the representativeness of NMTBA policy positions. A brief examination of the association's Board of Directors addresses and assuages this issue. Table 3.7 lists the ten NMTBA directors for 1982 and their 1980 machine tool sales.

The board is very balanced in terms of the sizes of the firms represented. Five of the top ten firms (plus a subsidiary of one of the top ten already on the board) are represented: Acme-Cleveland, Bendix, Giddings & Lewis, Litton, and Textron. Four more directors are from medium-to-large builders: Brown & Sharpe, Minster, Oerlikon, and Verson. The final four directors are from (very) small members: Cincinnati Gilbert, Gallmeyer & Livingston, Rousselle, and Williams. The president of the smallest firm on the board, Gallmeyer & Livingston with sales of only $5 million, is the board's chairman.

Two interpretations can be made regarding the composition of the board and the policy positions that emanate from it. First, NMTBA positions can be considered to reflect a consensus, which is very possible in such a small industry, or at least the lowest common denominator of the entire industry. Alternatively, despite the formal composition of the board, the association's policymaking process may be dominated by the larger firms. In the first case, association public policy positions can be taken at face value as representative of the industry as a whole. In the second case, particular attention should and will be given to the large builders, particularly when seeking to explain (in Chapter IV) the trade policy views that follow. By making these latter adjustments, the usefulness of NMTBA policy positions is not invalidated, and no determination need be made regarding the relative validity of the two alternative interpretations.

Table 3.7

1982 National Machine Tool Builders' Association Board of Directors

	Firm	*1980 Sales ($m)*	*/ Sales Category*
#5	Acme-Cleveland	257	capital equipment
#2	Bendix, Industrial Equip. Group	623	industrial
	Brown & Sharpe Mfg. North Kingston, Rhode Island	93	machine tools
	Cincinnati Gilbert Machine Tool Cincinnati, Ohio	8	total
	Gallmeyer & Livingston Grand Rapids, Michigan (Chairman)	5 *	total
#6	Giddings & Lewis	222	machine tools
#10	Litton	190	machine tools
	Minster Machine Minster, Ohio	80	total
	Oerlikon Motch Cleveland, Ohio	156 **	total
	Rouselle Chicago, Illinois	NA	
	Sheldon Machine Chicago, Illinois	***	

(continued)

Table 3.7 (continued)

	Firm	*1980 Sales ($m) / Sales Category*	
#7	Textron, Bridgeport Machines Div.	250 *	machine tools
	Verson Allsteel Press Chicago, Illinois	100	total
	Williams, White Moline, Illinois	14	total

* Estimate
** Sales for Motch & Merryweather Machinery only
*** Subsidiary of Acme-Cleveland
NA Not Available

Rank in Top Ten as indicated

Sources: *American Machinist*, August 1981, pp. 59-60; *Annual Reports*; Dun & Bradstreet, Dun's Marketing Services, *Million Dollar Directory 1982.*

1950-1969

The industry's first postwar concerns over import trends were expressed in reaction to the first major import surge of 1950-1952.[79] Complaints about cheap European imports were voiced in Congress in 1952, 1954, and 1955.[80] In 1955, the NMTBA issued a major position paper entitled, "The Need for a National Policy for the Machine-Tool Industry in Peace and War."[81] In a major section on trade policy, the association recommended "that a major effort should be the negotiation of future reciprocal trade agreements to obtain concessions regarding machine tool tariffs and other trade barriers from the principle customer countries." After noting that the balance of machine tool restrictions favored foreign competitors, the association observed that "corresponding concessions from this country need not take the form of a reduction in United States machine tool tariffs, at least in the first instance." The NMTBA went on to propose that

> meanwhile and in the event that imports of foreign machine tools increase above their 1954 level, it may be necessary or advisable for the President to raise tariffs or impose import quotas. . . . In this way . . . it may ultimately be possible to produce a proper balance between foreign and domestic tariffs and other restrictions in international machine tool trade.[82]

From a minimal position of opposition to further tariff reduction, the NMTBA moved a step closer toward seeking heightened trade protection in 1958. After noting U.S. postwar assistance in rebuilding the European and Japanese machine tool industries, J. Herbert Meyers of the association raised before the House Ways and Means Committee:

> the question whether the national interest does not require some measure of protection against the foreign machine tool competition which the United States created.[83]

Meyers recommended the strengthening of the protective features of the Trade Agreements Act of 1955. After observing that the President was authorized to raise tariffs by only 50%, Meyers announced

> it is the position of the machine tool industry that the President should have clear authority to raise the tariff as high as is necessary in the interests of national security.[84]

In specific reference to machine tools, Meyers felt that the industry deserved such protection:

> We feel that increasing the tariff is a proper remedy provided it can be increased to a point where the imports of foreign machine tools which are now at considerably lower prices are not going to come in at such prices as to close out those companies who are presently producing those machines.[85]

During hearings on the Trade Expansion Act of 1962, the association returned to its minimal position. Francis J. Trecker, President of Kearney & Trecker and NMTBA President, urged the retention of existing U.S. machine tool tariffs. Accordingly, the association opposed the Trade Expansion Act which permitted the President to reduce machine tool tariffs to zero. Trecker proposed alternative legislation whereby Congress would list those industries considered vital to national security. Consent by the Tariff Commission and Office of Emergency Planning would be required for the President to reduce tariffs on products of these industries. And, of course, the "machine tool industry should be high on that list of essential industries." The association went on to stipulate that tariff concessions should remain on an industry-by-industry basis and not be made across-the-board, and that Presidential tariff reduction authority be reduced from a five-year to a two-year period.[86]

The association's opposition to negotiated tariff reductions was reaffirmed in 1964 and 1968. Yet in direct reaction to the second major import surge of the mid-1960s, the association moved to its most protectionist policy stance in the postwar period. In testimony before the House Ways and Means Committee in 1968, Philip O. Grier, President of Cincinnati Milling Machine (since renamed, Cincinnati Milacron), Daniel W. Leblond, President of Leblond Machine Tool (Cincinnati, Ohio), and George W. Herkner, Executive Vice President of Warner & Swasey proposed an elaborate system of machine tool import surcharges. Incrementally higher duties would be imposed on levels of imports rising above 10% of domestic consumption for any or all machine tool product classes. At the time, import penetration levels for boring machines stood at 20%, drilling machines at 14%, milling machines at 12%, lathes at 11%, and all other metal-cutting tools at 8% (see Table 3.4). While not advocating quantitative limitations on imports, the NMTBA proposal would have added a 3% ad valorem surcharge for each import penetration percentage point exceeding 10% up to 15%; a 4% surcharge for each percentage point above 15%; with an absolute surcharge ceiling of 50% ad valorem. The system was to be implemented and phased out over a fifteen year period.[87]

1970-1981

In 1968, the machine tool industry had taken its furthest steps down the protectionist path. Two years later, however, the industry had already begun backtracking and by 1973 the industry would be taking its strongest free trade stance since the Second World War.

In 1970, again in hearings before the House Ways and Means Committee, Henry D. Sharpe, President of Brown & Sharpe and NMTBA First Vice President, and Daniel Leblond began to back off from the association's surcharge proposal. The industry still expressed grave concern over the import situation and still opposed negotiation of further U.S. tariff concessions. Yet the industry spokesmen offered the strengthening of escape clause provisions of U.S. trade law as an alternative to its 1968 proposal. In seeking "the preservation of our vital machine tool production capacity," the association conceded that

> in recent months, however, we have concluded that a liberalized and revitalized Escape Clause might well be the least controversial and most practically available route for achieving that result.[88]

The escape clause provision is certainly a protectionist vehicle, but a much more generic, procedural one compared to a protectionist scheme designed specifically for machine tools and for immediate implementation. To help establish the significance of what may appear to be a minor policy change, this policy retreat was made at the very same time, indeed, at the very same congressional hearings when industry after industry were joining the protectionist bandwagon led by Congressman Wilbur Mills and his import quota legislation of 1970.

By 1973 the industry consummated the policy reversal begun in 1970. In 1973 and 1974 the association, as represented by Philip Grier and Ralph E. Cross, President of Cross Company and NMTBA First Vice President, voiced strong *support* for the Trade Reform Act of 1974. In line with its 1970 position, the association welcomed the strengthening of escape clause provisions. Yet in a postwar precedent, the industry also vigorously backed the Act's provisions to "confer upon the President broad authority to negotiate reductions in such U.S. tariffs as survived the Kennedy Round negotiations of the 1960s or in some cases to eliminate them altogether."[89] Grier, Cincinnati Milling Machine President, then established what would become the industry's general trade policy for the 1970s:

> While NMTBA is concerned over the inroads into the U.S. market made by foreign competition, particularly in certain machine tool lines, it believes that in the long run free trade — assuming

> reciprocity, fair trading practices on all sides and adequate governmental authority to deal with emergency situations — is both inevitable and desirable.[90]

In direct opposition to NMTBA's 1968 trade policy position, Cross emphasized that

> we do not favor general tariff increases or the widespread imposition of quantitative restrictions on imports.[91]

These positions are exceptional and not only in comparison to the industry's prior trade policy views. The advocacy of free trade is also striking in light of the political and international economic environment of 1973-74. Politically this was again a time where there was an industry-wide protectionist movement, this time behind the Burke-Hartke trade legislation. And economically, the industry had more reason to seek protection that at any time prior. Between 1968, the year the NMTBA made its import surcharge proposal, and 1974, machine tool imports rose a colossal 80%.[92] And in 1974 nearly 20% of the U.S. market was captured by these imports compared to under 12% in 1968. It is at this point that the anomalous behavior of the U.S. machine tool industry comes into clearest relief. Through the 1970s and into the 1980s, the industry reaffirmed its free trade advocacy as evidenced in the following public statements:

> The U.S. machine tool industry has been an outspoken advocate of free trade. We have steadfastly supported the principle of free and fair competition among the machine tool builders of the world. And, we still firmly believe in free trade. (Testimony by James A. Gray, NMTBA President, at House Ways and Means hearings on the Causes and Consequences of the U.S. Trade Deficit and Developing Problems in U.S. Exports, November 3, 1977, p. 393.)
>
> We are, of course, aware that many in U.S. industry have called for import quotas on such Japanese goods in the interest of protecting U.S. workers from "foreign-born" unemployment. Although we

> certainly sympathize with the objective of keeping as many U.S. workers employed as possible, we must disagree with restrictive import quotas as the method of achieving the goal. (Testimony by Frank J. Austin, Executive Vice President and Chief Executive Officer of Giddings & Lewis, at House Ways and Means Subcommittee on Trade Hearings on Trade with Japan, August 26, 1980, p. 167.)
>
> Because the U.S. machine tool industry continues to maintain a preference for free and fair trade, we fully concur in the Secretary's [Secretary of Commerce Malcolm Baldridge] belief that the current U.S.-Japanese trade imbalance should be "redressed by expanded U.S. exports to Japan rather than artificially reduced Japanese exports to the U.S. market." (Statement by the NMTBA before the House Ways and Means Subcommittee on Trade, December 15, 1981, p. 21.)

The industry gave broad support to the Tokyo Round multilateral trade agreement of 1979. It was pleased with the negotiated mutual tariff reductions believing that a relative balance of concessions given and received had been achieved. The industry supported both the agreement's subsidy code and its government procurement code.[93]

The machine tool industry did not, however, blindly follow a free trade philosophy. From the 1970s, the industry was keen to point out the lack of reciprocity in foreign trade practices. In particular, machine tool builders made public complaints regarding government-sponsored export financing and other industrial subsidies, patent infringements, and circumvention of COCOM export controls.

Yet the most serious of such charges levied by the industry concerned alleged dumping, primarily on the part of Japan, during 1977-1978. Beginning in the fall of 1977, a growing number of complaints regarding excessively low-priced Japanese tools were made to the NMTBA by member companies.[94] In November, NMTBA President Gray spoke before the association's annual meeting and the Ways and Means Subcommittee on Trade charging "apparently illegal and predatory practices of foreign builders and their governments."[95] Gray cited instances where Japanese machines were marketed in the U.S. for $360,000 less than in Canada, and where Japanese tools were selling for 50% of the price of comparable U.S. machines. He went on to declare:

> There is a strong odor here . . . . It smells like dumping, and you may be assured that NMTBA will follow the scent until it is clearly identified and neutralized . . . . If the facts show that the Japanese, or any other foreign builders, are dumping their machines in the U.S., then NMTBA will spend whatever resources necessary to bring it to a halt.[96]

The association then surveyed its membership for reports of foreign dumping and distributed bulletins outlining the evidence necessary to file a dumping suit. The NMTBA even appointed a special committee composed of leading machine tool firms to study the problem and seek simplified procedures for bringing dumping charges.[97]

Dumping charges may be legitimate responses to imports sold at less than fair value (i.e., export prices below domestic prices in the producing country). In this case, no protectionist connotations can be imbued upon those bringing dumping charges. On the other hand, as discussed in Chapter II, dumping allegations may be levied to harass foreign suppliers and otherwise restrain imports on grounds other than unfair pricing.

In the machine tool case, dumping charges were dropped before formal adjudication so the merits of the NMTBA allegations were not determined. Yet insight into the industry's motivations is provided by the reasons the NMTBA dropped its dumping actions. In brief, the charges were terminated when price differentials between Japanese and U.S. machines dissipated in early 1978. Fueled in part by the depreciation of the dollar against the yen (see earlier discussion of price-based Japanese competition) and by efforts on the part of Japanese machine tool builders,[98] prices of Japanese exports to the U.S. rose between 20% and 25% in early 1978.

The fact that the dumping motions were tossed out with these developments reveals that price considerations and not generally protectionist aims motivated the industry. Such an interpretation is all the more supported by the fact that despite the price developments, Japanese exports to the U.S. grew at a faster rate after 1977 than before (1977-1979 annual rate of 85% versus 33% for 1975-1977).[99] If concerns about the level of Japanese imports motivated the industry's dumping charges, then the complaints should have heightened, not disappeared, after 1977.

This interpretation is also consistent with the measured behavior of the industry even when it was concerned about foreign dumping. The industry prepared the groundwork to file antidumping suits, but did not rail against foreign imports and foreign trade generally. Unlike many other industries that have made charges of foreign dumping, the machine tool builders did not exploit the issue to seek wider protective measures. Instead, at the time the dumping issue was raised in 1977-78, the industry was advocating its support for free trade and endorsing the Tokyo Round tariff reduction negotiations.

In summary, the U.S. machine tool industry was not without its criticism of foreign trade relations and U.S. import trends. No realistic observer could expect let alone postulate otherwise. Yet what is surprising is not that such trade complaints were made, but that they were so restrained. In comparison to large segments of U.S. industry, machine tool builders can accurately be characterized as having taken relatively liberal commercial policy stances. And all in the context, analyzed in the first half of the chapter, of heightened foreign competition and the loss of larger and larger as well as more significant and sophisticated segments of the U.S. machine tool market to foreign producers. While the industry shifted from its most protectionist period in the late 1960s to its liberal commercial policy positions from the mid-1970s, machine tool import penetration levels rose from 12% to 30%. This, of course, is our anomaly for which an explanation is sought in the next chapter.

NOTES

[1] U.S., Bureau of the Census, *U.S. Commodity Exports and Imports as Related to Output,* ES2; *U.S. Exports. Domestic Merchandise, SIC-Based Products by World Areas*, FT 610; U.S. Bureau of Economic Analysis, *Business Statistics*; *Survey of Current Business.*

[2] U.S., Department of Commerce, International Trade Administration, *International Economic Indicators*; U.S., Bureau of Economic Analysis, *Survey of Current Business*.

[3] U.S. Bureau of the Census, *U.S. Commodity Exports and Imports*:; *U.S. Imports: Consumption and General, SIC-Based Products by World Areas*, FT 210; U.S., Bureau of Economic Analysis, *Business Statistics; Survey of Current Business*.

[4] See footnote no. 2.

[5] *Washington Post,* May 11, 1982, p. A8.

[6] For a description of these product types see Geoffrey Boothroyd, *Fundamentals of Metal Machining and Machine Tools* (New York: McGraw Hill, 1975).

[7] Unless otherwise indicated, this and the following paragraph is based on U.S., International Trade Commission, *Summary of Trade and Tariff Information: Metalworking Machine Tools*, May, 1981, p. 3.

[8] Dexter S. Kimball, at the time Dean of Engineering at Cornell University, coined this often used phrase. See U.S., Congress, Senate, Committee on Finance, *Trade Expansion Act of 1962*, hearings, part 3, August 7, 1962, 87th Congress, Second Session, p. 1605.

[9] U.S. International Trade Commission, *Summary of Trade*, p. 3.

[10] Cited in U.S., Congress, House, Committee on Ways and Means, *Foreign Trade and Tariff Proposals*, hearings, part 7, June 21, 1968, 90th Congress, Second Session, p. 2852.

[11] Projected from National Machine Tool Builders Association (NMTBA), *1981-1982 Economic Handbook of the Machine Tool Industry*, p. 4.

[12] James A. Gray, "The American Machine Tool Industry", February 12, 1982, p. 2 (mimeo).

[13] Cited in U.S., International Trade Commission, *Summary of Trade*, p. 9.

[14] Gray, "The American Machine Tool Industry," p. 2.

[15] Figures in this paragraph from U.S., Bureau of the Census, *1977 Enterprise Statistics: General Report on Industrial Organization*, April 1981.

[16] U.S., Bureau of the Census, *1977 Census of Manufactures: Industry Series, Metalworking Machinery and Equipment*, June 1980.

[17] U.S., Comptroller General, "Followup on Use of Numerically Controlled Equipment to Improve Defense Plant Productivity," LCD-78-427, January 11, 1979, p. 6 cited in Judith Reppy, "The Role of the Air Force in the Development of Numerically-Controlled Machine Tools," 1981 (mimeo).

[18] Unless otherwise indicated, all import penetration figures are based on monetary value. The alternative is a unit (quantity) base.

[19] U.S., Congress, House, Committee on Ways and Means, *Trade Agreements Extension*, hearings, part 2, January 17, 1955, 84th Session, p. 3.

[20] U.S., Congress, House, Committee on Ways and Means, *Trade Expansion Act of 1962*, hearings, part 2, March 12, 1962, 87th Congress, Second Session, p. 14; U.S., Congress, Senate, Committee on Finance, *Trade Expansion Act of 1962*, hearings, part 3, August 7, 1962, 87th Congress, Second Session, p. 5.

[21] U.S., Bureau of the Census, *U.S. Imports for Consumption: TSUSA Schedule by TSUSA Number by Unit Control by Country of Origin*, IM 146.

[22] U.S., Congress, *Foreign Trade and Tariff Proposals*, p. 2858.

[23] *Steel*, December 10, 1962, p. 31; U.S. Congress, *Foreign Trade and Tariff Proposals*, pp. 2858-60.

[24] U.S., Bureau of the Census, *U.S. Imports for Consumption: TSUSA*; NMTBA.

[25] U.S., Bureau of the Census, *U.S. Exports: Schedule B Section by Division by Group by Number by Country of Destination by Customs District of*

Exportation and Method of Transportation, EM 522; *U.S. Imports for Consumption: TSUSA.*

[26] *Iron Age*, April 14, 1980, p. 41; *American Metal Market/Metalworking News*, November 23, 1981, p. 19.

[27] All figures from U.S., Bureau of the Census, *U.S. Imports for Consumption: TSUSA*; NMTBA.

[28] Japan, Tariff Association; Ministry of Finance.

[29] Japan, Tariff Association; U.S., Bureau of the Census, *U.S. Exports: Schedule B*; *U.S. Imports for Consumption: TSUSA*; "Current Industrial Reports, Series MQ-35W, Metalworking Machinery." Since Japan's export figures are based on Japanese sources, the resulting import penetration ratios are not directly comparable to U.S.-sourced ratios.

[30] Japan, Tariff Association; Ministry of Finance.

[31] Statement by National Machine Tool Builders' Association before the Subcommittee on Trade, Committee on Ways and Means, U.S. House of Representatives, December 15, 1981, p. 5 (mimeo).

[32] Japan, Tariff Association.

[33] *Industry Week*, May 18, 1981, p. 71.

[34] For a discussion of these two terms see Wilbur F. Monroe, *Japanese Exports to the United States: Analysis of Import Pull and Export Push Factors* (Washington: U.S.-Japan Trade Council, June 2, 1978).

[35] *Iron Age*, May 3, 1973, p. 47.

[36] *Standard and Poor's Industry Surveys*, October 1981; *Iron Age*, January 15, 1979.

[37] *Industry Week*, September 1, 1980, p. 50; *Iron Age*, February 23, 1981.

[38] *Iron Age*, February 23, 1981.

[39] *New York Times*, April 26, 1982, p. Dl.

[40] U.S., Congress, House, Committee on Ways and Means, *Tariff Readjustment - 1929*, hearings, v. 3, schedule 3, January 14. 1929, 70th Congress, Second Session, p. 1.

[41] The following section contains such examples.

[42] Computed from U.S., Bureau of the Census, *U.S. Imports for Consumption: TSUSA*; "Current Industrial Reports."

[43] *Industry Week*, September 16, 1974.

[44] *American Metal Market/Metalworking News*, July 17, 1978, p. 24; September 18, 1978, p. 15; *Iron Age*, September 18, 1978, p. 12.

[45] *Iron Age*, September 18, 1978, p. 12.

[46] Monroe, *Japanese Exports to the United States*, p. 165.

[47] *American Machinist*, June 1981, p. 61.

[48] *American Metal Market/Metalworking News*, June 30, 1981, p. 27.

[49] *American Machinist*, November 1980, p. 61.

[50] *American Metal Market/Metalworking News*, February 22, 1982.

[51] See "Machine Tools: Losing that Cutting Edge," *Industry Week*, September 1, 1980; "Are Machine Tool Builders Cutting it with Industry?" *Iron Age*, August 24, 1981; *Iron Age*, August 21, 1978, p. 15.

[52] *American Metal Market/Metalworking News*, February 22, 1982.

[53] U.S., Congress, House, Committee on Ways and Means, Subcommittee on Trade, *Causes and Consequences of the U.S. Trade Deficit and Developing Problems in U.S. Exports*, hearings, November 11, 1977, 95th Congress, First Session, p. 388.

[54] U.S., Bureau of the Census, *U.S. Exports: Schedule B*; *U.S. Imports for Consumption: TSUSA*; "Current Industrial Reports."

[55] Houdaille Industries, Inc. *Petition to the President of the United States Through the Office of the United States Trade Representative for the Exercise*

of Presidential Discretion Authorized by Section 103 of the Revenue Act of 1971, 26 U.S.C. 48(a)(7)(D), May 3, 1982, p. 124.

56 *American Metal Market/Metalworking News*, May 18, 1981, pp. 28-29.

57 *Ibid.*, December 21, 1981, p. 10.

58 *Iron Age*, September 15, 1980.

59 *Ibid.*, September 28, 1981.

60 *Ibid.*, October 15, 1980.

61 *Industry Week*, May 18, 1981.

62 *American Metal Market/Metalworking News*, July 27, 1981, p. 12.

63 *Ibid.*, June 30, 1980, p. 26.

64 *Industry Week*, February 9, 1981, pp. 65-58.

65 *Ibid.*, May 18, 1981.

66 *American Metal Market/Metalworking News*, January 19, 1978, p. 1.

67 *Iron Age*, August 27, 1979, pp. 101-2.

68 *Ibid.*, September 28, 1981, p. 88.

69 *American Metal Market/Metalworking News*, November 30, 1981, p. 12a.

70 *Ibid.*, January 11, 1982, p. 12.

71 *Iron Age*, November 2, 1981.

72 *Iron Age*, September 28, 1981; American Machinist, March 1980, p. 181.

73 See, for instance, Clausing Machine Tool's (Elgin, Illinois) experiences, *Iron Age*, May 21, 1979, p. 19.

[74] Import penetration levels for footwear and mushrooms, since initial petitions for import relief, grew to those for machine tools.

[75] U.S. Congress, House, Select Committee on Export Control, *Investigation and Study of the Administration, Operation, and Enforcement of the Export Control Act of 1949, and Related Acts*, hearings, November 25, 1961, 87th Congress, First Session.

[76] U.S., Congress, Senate, Banking and Currency, *Export-Import Bank Act Amendments of 1954*, hearings, June 14, 1954, 83rd Congress, Second Session; U.S., Congress, Senate, Committee on Banking, Housing and Urban Affairs, Subcommittee on International Finance, *Export Policy*, hearings, part 4, March 20, 1978, 95th Congress, Second Session.

[77] See Robert O. Keohane and Joseph S. Nye, *Power and Interdependence: World Politics in Transition* (Boston: Little, Brown, 1977): Stephen D. Krasner, "The Tokyo Round: Particularistic Interests and Prospects for Stability in the Global Trading System," *International Studies Quarterly*, v. 23, no. 4 (December 1979); Susan Strange, "The Management of Surplus Capacity: Or How Does Theory Stand Up to Protectionism 1970s Style?" *International Organization*, v. 33, no. 3 (Summer 1979).

[78] "Statement by Wayne R. Moore, President and Chief Executive Officer, Moore Special Tool Company, Inc., before the Subcommittee on Financial Institutions, Supervision, Regulation and Insurance, Committee on Banking, Housing and Urban Affairs, U.S. House of Representatives, April 22, 1982," p. 5 (mimeo). Since at least 1952, the association has represented at least 80% of U.S. machine tool production.

[79] The remainder of this chapter refers to import trends analyzed in the first part of the chapter.

[80] U.S., Congress, Senate, Select Committee on Small Business, *Machine Tool Shortages*, hearings, February 27, 1952, 82nd Congress, Second Session; U.S. Congress, *Export-Import Bank Act*, June 14, 1954; U.S., Congress, *Trade Agreements Extension*, January 17, 1955.

[81] Included in U.S., Congress, Senate, Select Committee on Small Business, *Machine Tool Programs*, hearings, February 7, 1956, 84th Congress, Second Session.

[82] *Ibid.*, p. 80.

[83] U. S., Congress, House, Committee on Ways and Means, *Renewal of the Trade Agreements Act*, hearings, part 2, March 10, 1958, 85th Congress, Second Session, p. 1744.

[84] *Ibid.*

[85] *Ibid.*, p. 1748.

[86] U.S., Congress, *Trade Expansion Act of 1962*, March 12, 1962.

[87] U.S., Congress, *Foreign Trade and Tariff Proposals*, June 21, 1968.

[88] U.S., Congress, House, Committee on Ways and Means, *Tariff and Trade Proposals,* hearings, part 9, June 4, 1970, 91st Congress, Second Session, pp. 2538-39; also pp. 2493, 2553.

[89] U.S., Congress, Senate, Committee on Finance, *The Trade Reform Act of 1973*, hearings, part 3, March 26, 1974, 93rd Congress, Second Session, p. 1505; see also U.S., Congress, House, Committee on Ways and Means, *Trade Reform*, hearings, part 3, May 14, 1973, 93rd Congress, First Session.

[90] U.S., Congress, *Trade Reform*, May 14, 1973, p. 803.

[91] U.S., Congress, *The Trade Reform Act of 1973*, March 26, 1974, p. 1506.

[92] U.S., Bureau of the Census, *U.S. Imports for Consumption: TSUSA*..

[93] "Report of the Industry Sector Advisory Committee on Machine Tools, Other Metalworking Equipment and Other Nonelectrical Machinery (ISAC 17) to the President, the Congress, and the Special Representative for Trade Negotiations, April 18, 1979," in U.S., Senate, Committee on Finance, Subcommittee on International Trade, *Private Advisory Committee Reports on the Tokyo Round of Multilateral Trade Negotiations*, August 1979, pp. 316-333.

[94] *American Metal Market/Metalworking News*, October 10, 1977, p. 2.

[95] U.S., Congress, *Causes and Consequences of the U.S. Trade Deficit*, November 31, 1977, p. 393.

[96] *Ibid.*, pp. 394-5.

97 *American Machinist*, December 1977, pp. 41-3.

98 Beginning in March 1978, but planned since at least July 1977, the Japanese Ministry of International Trade and Industry (MITI) established minimum f.o.b. prices for NC lathe and machining center shipments to the U.S. Administered by the Japan Machinery Exports Association and the Machine Tool Exporters' Association, exports have been screened to ensure that machine prices do not fall below set floor prices. This export price cartel has been renewed annually.

In an important misinterpretation, the metalworking press reported the formation of the cartel as a direct response to defuse U.S. builder charges of dumping. (See *American Machinist*, January 1978, p. 45; May 1978, p. 45; January 1981, p. 47.) In point of fact, as reported in interviews with NMTBA officials, the Japanese cartel was planned in July 1977, if not earlier, well before Gray made his November speeches. The cartel was formed by MITI to give some order to the fierce competition (e.g., price discounting) among Japanese machine tool builders for market shares in the U.S., and was not designed to allay U.S. machine tool builder concerns nor to dampen exports to the U.S. Indeed, between 1978 and 1979, U.S. imports of Japanese NC lathes rose 93% and for machining centers 125% (*Iron Age*, March 17, 1980). The unilateral nature of the formation of the Japanese cartel was significantly confirmed by a U.S. Justice Department antitrust investigation. After three years of inquiry, the case was dropped when Justice could find no evidence that NMTBA actions prompted the Japanese cartel (*American Metal Market/Metalworking News*, October 2, 1978, p. 17.)

99 U.S., Bureau of the Census, *U.S. Imports for Consumption: TSUSA; U.S. Exports: Schedule B.*

Chapter IV

Explaining U.S. Machine Tool Trade Policy

INTRODUCTION

Our "problem" to be explained can now be stated in stark terms: Despite severe import penetration and hegemonic decline, the U.S. machine tool industry maintained relatively liberal trade policy positions through the 1970s. Mainstream work in international political economy, which seeks to explain trade protection in terms of the vissitudes of foreign imports and/or hegemonic decline, would dub the machine tool industry as classically ripe for if not rife with protectionism. Add on the strategic military and economic value of the industry and machine tool protectionism should be a political given. Yet, in Chapter III, we found this not to be the case for the 1970s. The task of this chapter is to seek to explain why not.

Before offering export dependence as an explanation, the following section considers and then rejects what would be three of the most prominent and conventional alternative explanations for machine tool trade policy: foreign direct investment, economic performance, and high technology. After concluding that our anomaly remains "unresolved," the next section goes on to make the case for the export dependence thesis. Given the constraints imposed upon protectionist behavior by export dependence, the chapter then outlines the machine tool industry's non-protectionist responses to foreign competition. These responses, both in the public policy and business policy realms, represent some of the "alternatives to protection" that could be undertaken by other export dependent industries facing severe international competition. The final section concludes with a brief discussion of developments in U.S. machine tool trade relations in the 1980s.

CONVENTIONAL EXPLANATIONS

In the aggregate data analysis of Chapter II, we explored the possibility that industrial trade policy could be explained by factors other than export dependence. A similar task is undertaken here for machine tool trade policy. In attempting to account for the industry's anomalous liberal commercial policy positions of the 1970s, three conventional explanations would be prominently offered involving foreign direct investment, economic performance, and technology.

Foreign Direct Investment

An industry's direct investments abroad could very well be responsible for the adoption of liberal commercial policy views. Such cannot be the case, however, for the machine tool industry because U.S. builders have not invested abroad heavily.

Beginning in the early sixties a few of the larger machine tool firms established multinational operations, predominately in Western Europe. Yet in comparison to other industries, these overseas facilities amounted to very little. In our aggregate data analysis for 1976 (see Chapter II), the metalworking sector ranked only 69th out of 129 sectors on foreign direct investment. Metalworking's ratio of dividends of foreign corporations and foreign tax credits to total assets amounted to only 0.6% — well below the national average of 1.4%. Table 4.1 illustrates the low levels of revenues provided and assets represented by the foreign operations of five leading machine tool builders that make such figures public.

Table 4.1

Foreign Operation Ratios of Leading Machine Tool Builders (%)

	1970	1971	1972	1973	1974	1975	1976	1977	1978	1979	1980	1981
Acme-Cleveland:												
Sales*	22.4	26.5	20.8	23.1					10.8	8.4	10.7	12.2
Assets**									10.1	1.3	1.7	1.7
Cincinnati Milicron:												
Sales					28.8	31.5		23.2	19.6	19.8	21.8	15.2
Assets		14.7	14.2	11.2	12.5	15.5		16.6	16.2	29.5	28.0	22.3
Cross & Trecker:												
Sales									18.9	14.8	17.1	14.8
Assets									23.5	20.5	17.4	14.0
Ex-Cell-O:												
Sales										10.7	13.1	11.1
Assets										9.1	11.0	10.9

(continued)

Table 4.1 (continued)

	1970	1971	1972	1973	1974	1975	1976	1977	1978	1979	1980	1981
Giddings & Lewis:												
Sales			13.7	14.8	12.3	11.3	9.2	11.2	11.9	10.9	10.5	
Assets			5.4	5.2	6.3	6.7	6.9	15.0	15.6	16.7	13.7	

* ratio of foreign operation sales to total sales
** ratio of foreign operation assets to total assets

Sources: *Annual Reports*

As it turns out, the mid-seventies may have been the beginning of the end for U.S. machine tool investment overseas. Beginning in mid-1970s, virtually every major firm that expanded abroad a decade earlier began to liquidate those operations. For example, Cross terminated its joint venture in Japan in 1975. At the same time, Kearney & Trecker liquidated its Japanese operation and downsized its British subsidiary. Warner & Swasey slashed its European operations, and Cincinnati Milacron closed one of its two West German plants. In 1976 Milacron shut down its French subsidiary, and in 1978 closed its remaining West German facility and consolidated its other overseas operations. Acme-Cleveland's LaSalle Machine Tool also shut down its Italian subsidiary.[1] Table 4.1 confirms this retrenchment as particularly manifested in the trends in Acme-Cleveland's assets, Cincinnati Milacron's sales, and Cross & Trecker's assets. Much like the U.S. steel industry, U.S. machine tool production is nationally not multinationally based.[2]

High Technology

Given the inadequacy of a foreign direct investment thesis, conventional analysis could attempt to explain liberal trade policy in terms of technology intensity. This argument cannot be applied to the U.S. machine tool industry for two reasons: machine tool builders are not technology-intensive; and the industry lost its technological edge over foreign builders.

Through the seventies, machine tool builders, even the largest, could in no way be characterized as "high-tech" firms. R&D spending in the industry is extremely low; in the mid-1970s only 1.3% of industry sales were devoted to R&D compared to 5% for all machinery industries, 3% for all manufacturing industries, and 7-10% for the computer and semiconductor industries.[3] In my own aggregate data analysis (Chapter II) — using an R&D proxy of the ratio of scientists and engineers to total employees, the metalworking industry's ratio of 3.2% fell well below that of chemicals (7.7%), electronics (7.1%), and even non-electrical machinery in general (4.5%).

The low levels of resource commitment to R&D even among the leading machine tool builders is evidenced in Table 4.2. In line with these low commitments to R&D, the industry does not have a high reputation for innovation. Due in part to the weak capital positions of its small firms and the cyclical nature of the business, the industry has

Table 4.2

Research & Development Expenditure Ratios* of Leading Machine Tool Builders (%)

	1971	1972	1973	1974	1975	1976	1977	1978	1979	1980	1981
Cincinnati Milacron	4.1	3.7	3.0	2.6	2.9	2.9	2.8	2.6	2.4	2.8	3.3
Cross & Trecker								1.1	1.2	1.2	1.1
Cross		1.0	1.0								
Kearney & Trecker					0.5	2.9	2.4	1.5			
Ex-Cell-O					1.5	1.9			1.4		
Giddings & Lewis				2.2	1.8	1.3	1.2	0.9	0.7	0.6	
Warner & Swasey					2.1	2.4					

* ratio of R&D expenditures to total sales

Sources: *Annual Reports*

been traditionally conservative, both in financial as well as technological terms.[4] George J. Becker, Giddings & Lewis President, has observed, "I believe that there's very little basic research going on in the machine tool industry . . . Most of what is charged for research is product development." And W. Paul Cooper, Acme-Cleveland President, admitted, "The companies in the machine tool industry have not been able to do a comprehensive development job."[5]

As a result, machine tool buyers have often grown dissatisfied with the "currentness" of U.S. machine tool technology. The automobile industry in particular has criticized machine tool builders for not developing new methods of machining aimed at stepping-up car production.[6] Machine tool research has even fallen on the shoulders of users. Major automobile companies have established their own machine tool research labs, and have licensed their results to builders for development.[7]

Not only have machine tool builders not been technology intensive and innovative, the industry has also lost its technological edge over foreign competitors. Chapter III's discussion of significant imports of high quality, sophisticated machine tools already indicates that the U.S. monopoly over advanced machine tool technology has long been broken. A brief discussion of developments in several machine tool technologies will illustrate the loss of U.S. worldwide leadership.

Numerical Controls. John T. Parsons developed the first NC unit in 1949 under the sponsorship of the U.S. Air Force. Bendix introduced the first commercial NC unit in 1954, and the first numerically controlled machine tools were marketed in 1955 by Giddings & Lewis, Jones & Lamson (Springfield, Vermont), and Sundstrand (Belvidere, Illinois).

Through the sixties and into the seventies, the U.S. was the world's leading NC producer.[8] In 1970, NC tools accounted for 13% of U.S. machine tool production compared to less than 8% of Japanese production.[9] Yet by 1977, Japan overtook the U.S. as the leading producer of numerical controls, and supplied approximately 60% of the world market by the late 1970s.[10] By 1980 a full 50% of Japanese machine tool production was NC controlled compared to only 26% of U.S. production.[11]

Japan's NC leadership was established largely on the efforts of a single company, Fujitsu Fanuc. This one firm has captured approximately 50% of the world NC market.[12] Fanuc attained what the OECD has deemed a "technological lead" in NC through sustained basic research and product development that has resulted in a product of high performance, extremely reliable hardware, and software particularly suitable for machine tools.[13] Fanuc along with its frequent cross-national partner, Siemens of West Germany,[14] were in the forefront of basic research in sophisticated, microprocessor-based NC systems.[15]

A major developing technology for NC concerns direct numerical control (DNC). Under DNC many machine tools can be directed by a single computer instead of each tool being directed by its own control. DNC is considered critical for progress towards the automated factory. In 1980, the Machine Tool Task Force, an industry-government-academic study group sponsored by the U.S. Air Force, determined that a half dozen enterprises each in Japan and Western Europe have major DNC development programs along with approximately a dozen U.S. firms.[16]

Unmanned Machining and Turning Centers. Kearney & Trecker developed the first machining center in 1958. The "Milwaukee-Matic" could mill, drill, tap, and bore a workpiece on four sides with up to 31 tools. By 1980 more than 55 builders from a half dozen countries were exhibiting centers at the NMTBA's biennial International Machine Tool Show in Chicago.

The latest developments in turning (lathe) as well as machining center technology concerns unmanned capability. A machining center with an automatic loading device and with untended, extended run capability was first developed by Kearney & Trecker in the mid-seventies.[17] Yet Japanese and West European machine tool builders have since taken the initiative in unmanned technology; indeed automatic loading capability is now considered to be in greater use and more sophisticated in Europe than in the U.S.[18] Among the most sophisticated and innovative unmanned turning and machining centers are made by Emag Maschinenfabrik of West Germany and by Okuma Machinery, Yamazaki Machinery Works, Ikegai Iron Works, and Kitamura Machinery of Japan.[19]

Flexible Manufacturing Systems. Kearney & Trecker also developed the first FMS in 1958. Milling, drilling and boring machines were tied together with a workpiece handling device and the entire cell was NC controlled. Kearney & Trecker has remained a leading U.S. developer and producer of FMS along with Cincinnati Milacron and Bendix.

Significant public and private programs for the development and production of FMS have been undertaken in Italy, Sweden, Britain, France, East Germany and the Soviet Union. Yet the most significant challenge to the U.S. in the field comes, of course, from Japan and West Germany.

Toshiba Machine is believed to have built the most FMS in the world, and the Yamazaki Machining System (see Chapter III) is considered to have "some of the most innovative ideas in modular machine building seen in the industry today."[20]

Perhaps the most impressive achievements in the field have been made by Fujitsu Fanuc. At the base of Mt. Fuji, Fanuc has constructed the world's first factory capable of untended, round-the-clock operation. The $37 million effort is based on 29 FMS machining cells each combining NC machine tools, monitors, and industrial robots. All the machining cells plus an automated storage and retrieval system are directed by a central computer terminal which is, in turn, connected to the computer at Fanuc's suburban Tokyo headquarters. The automated facility produces turning machines, machining centers, robots, and electrical discharge machines.[21]

Since 1975, three MITI research institutes and 20 manufacturers of materials, machine tools, and controls (firms contributing 80% of expenditures) have undertaken a Flexible Manufacturing System Complex project designed "to make Japan a world leader in systems that can produce parts automatically in small quantity and in a wide variety of configurations."[22] A major U.S. trade journal has reported that "machine tool officials generally agree that the Japanese are ahead of the U.S. in some aspects of systems construction."[23]

One of the most striking findings of the Machine Tool Task Force, the U.S. Air Force-sponsored study group on machine tool technology in 1980, was that in FMS technology "West Germany may be ahead of the U.S. in some areas." This finding was largely based on observations of the Augsburg machining complex of the West German aerospace manufacturer, Messerschmitt-Boelkow-Blohm (MBB). The $50 million complex is equipped with 28 FMS machining cells and self-propelled robo-carriers for parts transfer to manufacture aluminum and titanium aircraft components. Vijay A. Tipnis, a noted machine tool technology consultant, has called the MBB facility "the finest flexible manufacturing system in the world."[24] Indeed, U.S. Air Force FMS efforts have been explicitly modeled on the MBB complex, and one of the Air Force's major contractors in this effort, Grumman, has hired MBB as a consultant and subcontractor.[25]

With just these few remarks it is clear that U.S. leadership in machine tool technology has seriously if not completely dissipated. U.S. machine tool builders have certainly fallen behind in terms of applied technology, machine tool design, and product innovation. Yet even in terms of basic research — in such fields as NC software, DNC, FMS, high speed machining, laser machining — the U.S. has come under serious challenge from West Germany and Japan, among others. In its 1979 technical survey of the world machine tool industry, the OECD concluded that "technologically, Japan is probably the most advanced country in the world."[26] A U.S. perspective is provided by Otto Kern, Ex-Cell-O Vice President: "The U.S. is at best on par, certainly not ahead of Germany and Japan in terms of advancements in equipment and reliability."[27] These conditions differ markedly from the 1960s when, as Eli S. Lustgarten, the Paine Webber investment analyst, put it, U.S. machine tool builders took "a great leap forward" over foreign competitors with the development of NC and other technological advances.[28]

In sum, machine tool firms have neither been technology-intensive nor highly innovative. In addition, the industry was most protectionist when its technological lead over foreign competitors was greatest (the 1960s), and maintained relatively liberal trade policy views while that lead diminished (the 1970s). A technological explanation for machine tool trade policy can be made only with great difficulty.

Economic Performance

In contrast to foreign direct investment and technology intensity, the state of machine tool economics would seem to have powerful influence over the industry's trade policy views. Indeed throughout this study, economic recession has been acknowledged as a "potent pressure" for protectionism along with import penetration. Yet an explanation of machine tool trade views based on economic performance also proves unsatisfactory.

Table 4.3 compares the profit-sales ratios of leading machine tool builders, the machine tool industry in general, and all manufacturers. Between 1969 and 1977, the machine tool industry performed well below the average profit ratios of all manufacturing industries. And, in general, the large machine tool builders have performed no better than the industry as a whole. In 1977, the unweighted profit-sales ratio for the six leading firms listed in Table

4.3 came to 4.9% which compares closely to the entire industry's level of 4.8%. The *Industry Week* and *Value Line* composites for top machine tool builders are also very similar to the industry average.

Machine tool firms, both large and small, suffered their worst financial straits between 1970 and 1972 when the industry went into the red. This period, as was discussed in Chapter III, was precisely when the industry made its transition to the adoption of liberal trade views — a fact devastating for any financial explanation of machine tool trade policy.

Table 4.3

Profit-Sales Ratios* for All Manufacturers, the Machine Tool Industry, and Leading Machine Tool Builders (%)

	1966	1967	1968	1969	1970	1971	1972	1973	1974	1975	1976	1977	1978	1979
Acme-Cleveland	10.1	8.3	8.1	7.2	5.0	-0.7	3.3	5.2	3.5	3.0	1.5	2.2	4.5	4.2
Cincinnati Milacron				5.1	5.7	0.7	0.7	2.6	2.3	2.2	2.2	3.9	5.2	7.8
Cross & Trecker										7.1	5.6	6.3	7.2	8.9
Cross				3.6	1.9	2.8	4.1	6.2	6.2	4.8	5.0	5.9	6.4	
Kearney & Trecker		7.9	6.7	4.4	0.2	-7.2	-42.2	1.1	6.8	9.3	6.3	6.7	8.4	
Ex-Cell-O	8.3	6.3	6.0	6.2	4.4	2.2	2.4	5.1	4.9	4.7	6.1	6.5	5.4	5.6
Giddings & Lewis	6.2	5.6	4.4	2.6	-3.5	-10.7	-0.7	1.3	5.1	4.8	3.2	5.5	9.2	11.2
Warner & Swasey	9.1	8.9	11.4	6.5	5.5	3.2	3.5	5.8	3.7	3.0	4.3	4.7	4.8	

(continued)

Table 4.3 (continued)

	1966	1967	1968	1969	1970	1971	1972	1973	1974	1975	1976	1977	1978	1979
Machine Tool Industry (NMTBA)	6.5	6.4	5.6	4.3	1.5	-1.7	-1.1	2.4	3.2	4.9	4.9	4.8		
Industry Week Leading Builders						2.6	3.3	4.9	4.1	3.5	4.0	4.9	5.8	7.8
Value Line Leading Builders												5.0	6.1	7.2
All Manufacturers (FTC)	5.6	5.0	5.1	4.8	4.0	4.1	4.3	4.7	5.5	4.6	5.4	5.3		
Industry Week All Manufacturers							5.1	6.0	5.3	4.9	5.5	5.4	5.5	5.6

* ratio of after-tax profits to total sales

Sources: Annual Reports; "Financial Analysis of Industry," *Industry Week*; U.S., Federal Trade Commission, "Quarterly Financial Report for Manufacturing Corporations;" *The Value Line Investment Survey*

On top of this dismal financial performance, the industry's productivity has literally stagnated over the past decade (see Figure 4.1). In constant 1967 dollars, real output per employee amounted to $13,000 in 1970 and $14,000 in 1981. Over the same period, Japanese productivity soared nearly 300% and now stands some five times higher than U.S. machine tool productivity.

Figure 4.1

Machine Tool Output per Employee
(1967 Dollars)

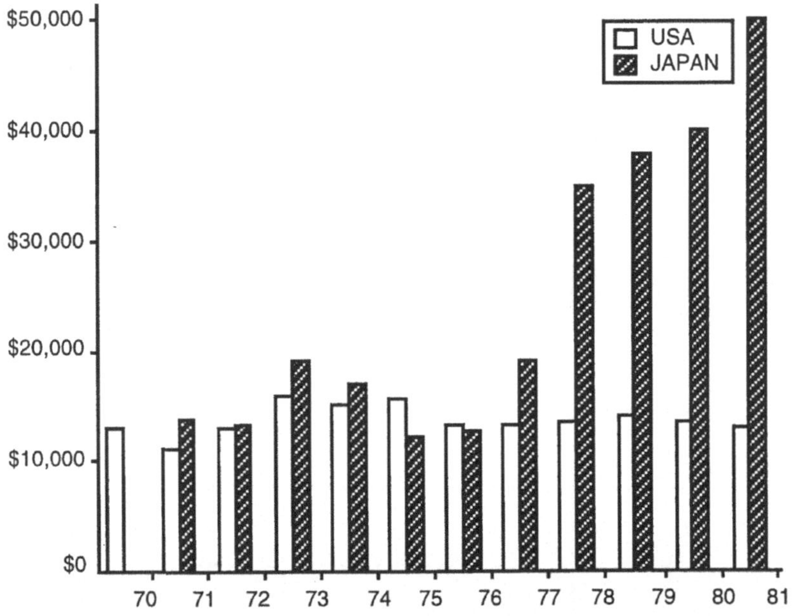

Sources: U.S. Department of Labor, Bureau of Labor Statistics, Japan Machine Tool Builders Association, NMTBA

The major yardstick by which the industry measures its own health is the level of orders for new machine tools. Machine tool orders are extremely sensitive to the state of the economy. The industry is usually the first to go into recession and the last to recover. Figure 4.2 displays the net new domestic orders for machine tools from 1956 to 1981 in current and constant prices. The distinguishing characteristic of the machine tool business is its boom and bust cyclicity.

Figure 4.2

Net New Domestic Orders for Machine Tools, Current and Constant (1967 Dollars)

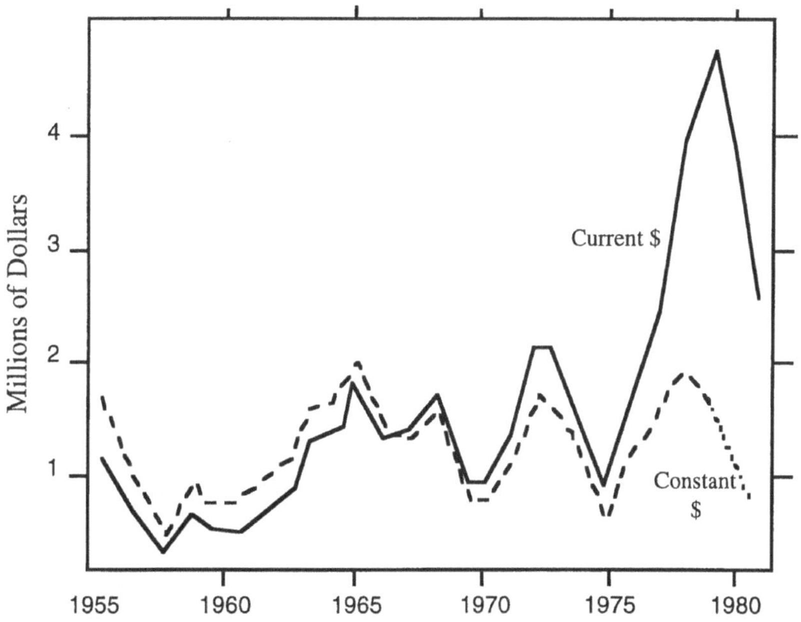

Sources: U.S. Bureau of Economic Analysis, *Business Statistics; Survey of Current Business;* U.S., Bureau of Labor Statistics, *Producer Prices & Price Indexes.*

Between 1956 and 1958, the industry took its steepest slide over the past 30 years. Industry backlogs got as bad as two months of work and machine tool capacity utilization fell as low as 30%. Needless to say, reduced work-weeks and layoffs were widespread.[29]

The industry subsequently experienced sustained recovery with orders gaining over 420% between 1958 and 1966 (all rates of change in real dollar terms). Business then slumped again with machine tool orders falling 63% between 1966 and 1971. Over this latter period machine tool employment shrank 35%.[30] And major firms were not spared earnings losses — including Brown & Sharpe, Cincinnati Milacron, Giddings & Lewis, Monarch, Warner & Swasey, Kearney & Trecker, and Acme-Cleveland.[31] Once again, it was during the depths of this machine tool slump that the industry began to backtrack from its protectionist position of 1968.

Industry orders surged in 1972-73 only to experience yet another dive in 1974-75. Order cancellations, which traditionally hover around 10% of gross orders, shot up as high as 50%. Heavy automotive cutbacks particularly hit the industry hard including the larger firms. In late 1974 and early 1975, General Motors and Chrysler canceled transfer lines for new front-wheel drive transmissions and small rack and pinion steering gear sets.[32]

Boom times returned in mid-1976, fueled largely by automobile industry orders for new machine tools. Responding to Federal fuel efficiency requirements and emission standards, the automobile makers began to retool to downsize and generally re-engineer its cars. This retooling was expected to have a long-term impact amounting to $60 billion for machine tools from 1980-1985 compared to the auto makers' $385 billion expenditures for tools between 1969 and 1978.[33] Combined with strong market demand by aerospace, construction, and energy industries and other manufacturers seeking higher productivity, *Standard & Poor's Industry Survey*, as late as 1981, put forward the "distinct possibility" that machine tool cyclicity could be erased with secular growth in the 1980s.[34] The Bureau of Industrial Economics of the Commerce Department predicted in its *1980 U.S. Industrial Outlook* that U.S. machine tool builders would enjoy annual growth rates of 9-10% through 1984.

This optimistic bubble began to leak in the second quarter of 1980 when construction firms and small shops began to hold off on machine tool purchases in response to the jump in the prime interest rate in March and April. By the end of 1980, larger industrial customers were delaying or reconsidering their tooling purchases as demand for durable goods plummeted.[35] In current dollars, 1980 domestic orders for new machine tools declined 14% from the prior year.[36]

By 1981 the bubble completely burst. Domestic orders fell to a five-year low — 40% below the already depressed level for 1980.[37] Industry backlogs fell to eight months of work from a 20 month high at the end of 1979.[38] In 1980 the smaller builders were hardest hit by the weakening of machine tool demand. By 1981 even the largest of firms were suffering due, in large part, to order developments in the industry's largest market: the automobile industry. Automobile manufacturers first delayed, then reconsidered, and finally canceled many tooling orders. These cutbacks involved not only Chrysler and Ford, but also General Motors — the company most expected would stand by its spending time-tables even if it meant borrowing heavily.[39] Through the year some $2 billion worth of orders were canceled by the car makers — the greatest loss of orders for a single year in machine tool history.[40] To top it off, the automobile makers made significant moves towards overseas sourcing for components and subassembly — further shrinking business for domestic tool builders.[41]

The machine tool recession has continued with full force into the first seven months of 1982. In March, industry backlogs had shrunk to only six months work.[42] Industry analysts predicted that annual orders would once again fall below the previous year. With industry capacity capable of handling $5 billion of shipments, 1982 orders were projected to amount to only $2 billion.[43] Cross, Bendix, LaSalle, Minster, and Ingersoll Milling Machine were among the industry giants hurt by this recession.

The machine tool business cycle would seem to have had significant impact on the industry's trade policy views. The economic slumps of 1956-58 and 1966-68 coincide with the growing protectionist sentiments of the late fifties and further hardening of trade policy views in the late sixties. The economic upswings of 1972-73 and 1976-79 also correlate with the liberal trade policy views of the 1970s.

While the impact of machine tool economics can never be denied, the explanatory power of the business cycle fails in a number of significant instances. First, protectionism flourished while the industry experienced sustained growth between 1958 and 1966. Economic growth, in this case, proved insufficient to sustain liberal trade policy.

More importantly, liberal policy — not protectionism — was embraced during the bust of 1971 and was maintained through the 1974-75 recession. These cases demonstrate that economic growth is not a necessary condition for liberal policy nor is economic recession sufficient to fuel protectionism. Something more than just the business cycle has been driving machine tool trade policy.[44]

To conclude this section, none of the three major conventional explanations of trade policy apply to the machine tool industry's liberal commercial policy views of the 1970s. A foreign direct investment explanation cannot work because the industry does not invest abroad heavily and has, in fact, been retracting its few foreign operations since the mid-seventies. A technological argument is unsatisfactory because the industry has not been technology-intensive and has lost its technological edge over foreign competitors. Finally, an economic performance thesis flounders with the finding that economic boom is not a necessary condition for liberal policy nor is economic bust sufficient to induce protectionism.

EXPORT DEPENDENCE

With the inadequacy of conventional explanations, our anomaly — liberal machine tool policy simultaneous with serious import penetration and hegemonic decline — remains "unresolved." Indeed, the industry's cyclical economic performance and relative technological decline make its commercial policy positions of the 1970s all the more anomalous. We are now in a position to consider the explanatory role played by the industry's dependence upon exports. Two questions, in particular, can be addressed:

1. Why was the industry protectionist through the 1950s and 1960s, and especially in 1968?

2. And why did the machine tool builders adopt liberal trade views in the early 1970s and maintain them through the decade?

The reference points for much of this discussion will be Figures 4.3 and 4.4 and Table 4.4 which present export dependence levels for all machine tools, metal-cutting machine tools, and seven product classes of metal-cutting machine tools. In each case three different export dependence series are presented. The first two correspond directly to the two import penetration series. The *product* series compares machine tool exports against machine tool shipments by all, primary and secondary, producers while the *industry* series is based on NMTBA shipments data. A third series, the *manufacturers* series, is based upon only the exports of manufacturers and not those of distributors and brokers. Unless otherwise indicated, reference to

Figure 4.3

Export Dependence*: All Machine Tools (%)

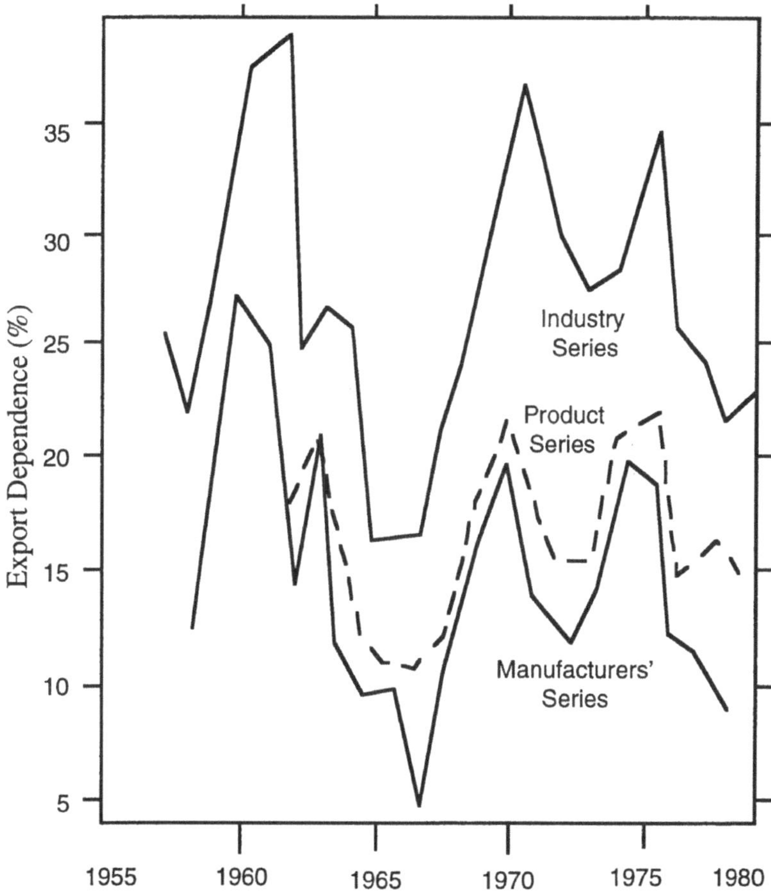

* ratio of imports to shipments

Sources: Industry Series – U.S., Bureau of Economic Analysis, *Business Statistics; Survey of Current Business*; U.S., Bureau of the Census, *U.S. Exports: Schedule B*, EM 522.

Product Series – U.S., Bureau of the Census, *Annual Survey of Manufactures: Value of Product Shipments; U.S. Exports: Domestic Merchandise, SIC-Based Products,* FT 610.

Manufacturer's Series – U.S. Bureau of the Census, "Current Industrial Reports, Series MQ-35W Metalworking Machinery."

Figure 4.4

Export Dependence: Metal-Cutting Machine Tools

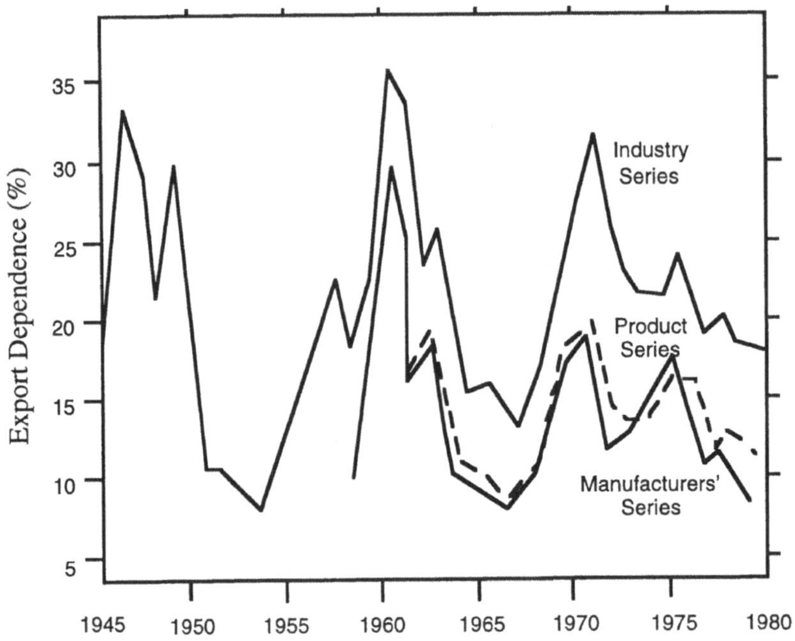

Source: see Figure 4.3

Table 4.4

Export Dependence: All Machine Tools, Metal-Cutting Machine Tools, and Metal-Cutting Product Classes (%)

(a) Industry Series

			Metal-Cutting Product Classes						
	All Machine Tools	Metal-Cutting Machine Tools	Boring Machines	Drilling Machines	Gear-Cutting Machines	Grinding Machines	Lathes	Milling Machines	Other Machines
1945		18.53							
1946		32.86							
1947		34.41							
1948		21.77							
1949		29.14							
1950		21.76							
1951		10.64							
1952		10.84							
1953		9.88							
1954		8.67							
1955									
1956									
1957									
1958	25.17	22.92	18.99	23.67	42.29	25.62	18.03	15.45	15.01
1959	21.92	18.66	13.64	10.78	45.34	18.76	14.56	11.60	13.22
1960	25.87	23.60	23.88	13.11	55.10	26.84	18.24	12.72	17.62
1961	35.98	35.40	26.80	19.02	71.70	43.14	31.94	21.71	30.09
1962	37.00	33.50	34.69	17.29	63.13	40.03	23.18	20.56	32.27
1963	24.63	22.70	16.79	12.12	36.46	27.27	16.86	10.74	25.27
1964	25.83	24.25	31.44	19.23	39.70	25.70	18.58	14.39	19.09
1965	20.15	18.74	12.95	10.34	28.90	17.51	10.01	5.21	19.94
1966	16.24	14.43	10.91	5.76	26.76	14.67	6.91	4.41	16.59
1967	16.31	14.75	11.04	8.88	36.93	12.91	6.47	4.82	16.90
1968	16.56	13.12	15.37	10.04	39.00	12.29	5.95	7.64	7.57
1969	20.25	16.86	14.59	14.07	37.20	18.26	7.68	7.37	8.98
1970	27.99	27.06	29.95	13.40	42.84	22.06	12.16	17.36	16.11

(continued)

Table 4.4 (continued)

(a) Industry Series (continued)

	Metal-Cutting Product Classes								
	All Machine Tools	Metal-Cutting Machine Tools	Boring Machines	Drilling Machines	Gear-Cutting Machines	Grinding Machines	Lathes	Milling Machines	Other Machines
1971	35.17	31.28	32.12	19.59	59.88	22.88	15.32	32.35	10.30
1972	32.60	26.13	20.70	14.44	59.64	20.00	10.38	16.65	8.65
1973	29.51	23.05	15.29	13.92	49.64	24.11	8.13	13.38	13.91
1974	27.90	22.19	17.34	16.56	43.96	21.21	8.37	10.75	16.71
1975	28.79	22.21	23.48	14.98	48.23	19.40	9.39	15.37	21.52
1976	33.15	24.47	20.95	13.55	45.89	23.22	11.02	10.71	18.13
1977	25.41	19.47	23.91	13.62	45.77	14.68	9.70	13.30	12.06
1978	24.51	20.73	18.44	9.06	45.76	23.18	8.15	15.75	13.09
1979	22.66	18.60	18.26	11.06	39.39	13.88	8.15	8.36	13.17
1980	23.43	18.33	17.19	11.90	46.97	15.12	8.11	5.66	11.93

(b) Product Series

1963	17.35	15.66	17.19	11.39	35.21	25.87	15.97	10.12	16.67
1964	19.45	18.04	30.38	18.96	39.04	24.55	17.71	14.26	17.64
1965	14.40	13.82	12.63	10.09	28.25	16.76	9.51	5.01	16.85
1966	11.63	10.82	10.49	5.52	26.38	14.13	6.54	4.40	12.02
1967	11.10	10.45	10.17	8.48	36.93	11.84	5.99	4.38	11.48
1968	11.09	9.79			38.83	10.87	5.71	7.01	5.66
1969	12.55	11.21			35.57	15.15		6.34	7.04
1970	16.94	17.40	18.81	8.96	37.52	20.44	11.82	12.2	15.12
1971	19.62	18.88	18.64	10.63	51.46	17.21	16.36	21.70	8.54
1972	16.34	15.08	19.08	12.89	52.91	17.91	10.68	15.70	7.19
1973	15.50	14.47	14.21	12.54	44.14	16.61	8.31	12.49	12.05
1974	15.64	15.02	16.53	15.91	41.45	19.64	7.83	10.33	14.36
1975	19.02	17.37	21.74	14.20	43.79	19.19	9.20	14.52	18.16
1976	19.57	16.90	20.32	13.74	41.52	23.53	10.54	9.81	15.74
1977	14.02	12.23	21.71	13.41	41.88	13.92	9.28	12.69	11.29
1978	15.40	13.60	16.32	8.90	42.15	22.48	7.89	14.94	12.18
1979	15.00	13.33	19.07	10.05	41.87	14.10	7.84	9.02	13.48

(continued)

Table 4.4 (continued)

(c) Manufacturers Series

	Metal-Cutting Product Classes								
	All Machine Tools	Metal-Cutting Machine Tools	Boring Machines	Drilling Machines	Gear-Cutting Machines	Grinding Machines	Lathes	Milling Machines	Other Machines
1959	11.89	10.56	7.79	2.99	44.13	13.75	8.54	6.96	7.45
1960	20.25	20.43	18.76	6.94	60.93	31.03	16.25	9.32	14.07
1961	27.10	28.61	24.08	10.05	62.64	38.60	25.06	14.38	27.87
1962	24.92	24.97	24.92	9.40	53.91	32.99	17.53	16.91	31.67
1963	14.93	15.53	11.34	9.04	34.43	22.08	13.70	7.54	16.62
1964	19.69	17.31	11.74	10.93	36.94	21.66	14.24	11.76	20.03
1965	11.75	12.13	5.22	5.05	33.39	14.56	8.56	6.11	17.58
1966	10.11	10.28	4.78	4.34	30.51	12.89	6.49	5.06	17.03
1967	10.38	10.48	4.07	5.86	46.01	11.99	5.60	7.17	14.81
1968	4.69	9.26	4.70	6.94	45.16	11.78	4.84	5.17	8.90
1969	10.73	10.88	7.73	8.30	49.09	16.28	5.82	3.63	9.35
1970	15.39	16.93	13.04	7.76	60.33	21.42	12.20	11.09	13.28
1971	18.06	18.64	10.61	11.52	72.65	20.57	13.72	21.69	10.72
1972	13.48	13.12	5.55	6.69	61.03	17.75	9.45	6.49	9.21
1973	12.57	13.90	5.55	6.63	53.35	18.81	6.79	7.65	15.99
1974	14.32	15.19	14.76	7.96	60.42	21.41	8.62	7.48	14.38
1975	17.61	18.15	9.65	10.02	55.15	18.12	11.85	15.37	21.67
1976	16.24	15.48	12.60	7.77	35.18	22.81	10.67	12.13	16.59
1977	11.38	11.58	7.66	3.92		14.02	8.16		9.94
1978	11.06	11.70	8.67	2.69		14.29	6.92		10.42
1979	9.33	9.19	5.36	2.79		8.84	7.00	7.18	

Sources: see Figure 3.1

specific export dependence levels will be taken from the industry series.

It will be recalled from Chapter III that the 1950s and 1960s were relatively protectionist times for the machine tool industry. On the basis of the export dependence thesis, one would hypothesize that these were times when the constraints of export dependence were low and/or diminishing. The historical evidence demonstrates such was the case for both decades.

1950-1969

In 1952 the industry began to lodge complaints against foreign import trends; complaints that resulted in the industry's opposition to tariff reductions beginning in 1955 and consideration of tariff increases in 1958. The adoption of these policy views coincide with deterioration in the industry's export performance in the 1950s.

In the immediate postwar period of 1946-1950, a full 28% of cutting tool shipments were exported to foreign markets (see Figure 4.4). During this period, Western Europe was a sellers' market for machine tools given the war-devastation of European machine tool capacity. In addition, early in this period substantial U.S. sales were made to the Soviet Union and Eastern Europe.[45]

All this changed with the onset of the Korean War. Industry efforts turned to domestic war requirements. By 1954, 91% of machine tool sales were for domestic consumption compared to 66% in 1947. Indeed, during the war, restrictions were placed on machine tool exports. At the same time Western Europe turned from a sellers' to a buyers' market. By the mid-1950s, European and especially West German machine tool industries were substantially rebuilt. NATO machine tool orders began to diminish as the alliance approached rearmament. With the dollar shortage, European imports of all but special machine tools were restricted by efforts to conserve hard currency. The Marshall Plan not only financed the rebuilding of European machine tool industries, among other industries, but deliberately reoriented European industrial orders from U.S. to indigenous manufacturers. Finally, machine tool sales to socialist countries were curtailed with the falling of the Iron Curtain.

Taken all together, major export markets in Europe contracted through the 1950s. The U.S. industry's export dependence ratio fell from 34% in 1947 to 9% in 1954. Metal-cutting exports fell another 35% between 1954 and 1959 (from $78 million to $51 million in current dollars.)[46] Whatever constraints that export dependence could impose on protectionist behavior were significantly diminished. And at the same time it will be recalled, the 1950s witnessed the first serious inroads into the U.S. market by foreign suppliers. Under these conditions of high motivation for and low constraints against protectionism, the industry's trade policy views of the 1950s can be understood.

By the early 1960s, the industry's export performance dramatically improved. Yet heightened levels of exports did not translate into effective constraints against protectionism. This seemingly incongruous development for the export dependence thesis occurred simply because the industry fully recognized at the time that its export boom was "without question a temporary condition."[47] The high level of machine tool exports in the early sixties was due "in large measure" to European machine tool builder backlogs of one to two-and-a-half years.[48] The growing European industry could not keep up with a boom in European machine tool demand, and U.S. builders filled this market gap with substantial export shipments.

Yet the export flurry was recognized as being the product of only "temporary circumstances."[49] U.S. builders knew full well that the growth of European machine tool capacity would soon meet if not exceed European demand. U.S. exports would plummet "just as soon as the rapidly expanding plant capacity of machine tool builders in the Common Market exceeds internal needs."[50] Hence, even during this export boom, U.S. machine tool builders were bracing themselves for an imminent loss of export markets — not to mention an invasion of their home market. Under these circumstances, it is not surprising that ephemeral export dependence levels did not heighten constraints against the industry's protectionism.

We have seen how conditions of high motivation for and low constraints against protectionism combined to harden the trade policy views of the machine tool industry in the late 1950s. Yet the classic case of high motivation and low constraint would take place in the late 1960s with the devastating result being the 1968 import surcharge proposal.

The fears of the early sixties were fully realized later in the decade when the bottom fell out for U.S. machine tool exports. Between 1961 and 1968, foreign orders for U.S. machine tools tumbled 35% in current dollars and 50% in constant dollars.[51] As a result, the industry's export dependence levels deteriorated dramatically. In 1962 a full 37% of industry shipments were exported. By 1968, less than 17% were shipped abroad. According to the manufacturers' series, export dependence fell from 27% to an incredibly low 5% (actually 4.7%) in 1968!

As shown in Figures 4.3 and 4.4, export dependence ratios have been quite cyclical for the machine tool industry. Yet the decline of the mid-to-late 1960s is notable for two reasons. The first concerns the steepness and the depth of the drop. According to the manufacturers' series, the level of export dependence plummeted 83% in an eight-year span. This decline exceeds the 75% drop in the eight years between 1947 and 1954. The export dependence valleys of the 1970s pale in comparison to that of 1961-1968. And the 4.7% export dependence level recorded in 1968 is by far the lowest level experienced in over 50 years.

The second significant aspect of the decline in the 1960s concerns its relationship to the machine tool business cycle. As one might expect, as domestic machine tool business sours the relative importance of export shipments (i.e., the export dependence ratio) increases. This phenomenon may be the result of machine tool firms (or any other firms) seeking out foreign buyers when the domestic economy is in recession. At the very least, the shrinking of domestic consumption during recession has the economic and statistical effect of heightening the relative position of exports. Conversely, when a domestic boom occurs the importance of exports declines. Again, this may either be the result of business policy (i.e., firms tending to growing domestic demand at the expense of foreign markets) or be a statistical artifact (i.e., inflating the denominator — shipments — in the export dependence ratio).

As demonstrated in Figure 4.5, the negative relationship between export dependence and order backlogs was maintained through the seventies and even in the early sixties. (It will be recalled that order backlogs is one of the industry's major indicators of the health of the business. Backlogs mount during good times and dissipate during bad.) When business boomed in 1972-74 and 1976-79, export ratios suffered. When business declined in the early sixties, 1971, and 1975-76, export dependence levels recovered.

Yet the one period that does *not* fit this pattern is that of the late sixties. Between 1966 and 1971 order backlogs shrank 69% in real terms, reaching its lowest point in 13 years. As discussed above, prior and subsequent recessions manifested greater reliance on exports. Business downturns, while heightening motives for protection, would almost have a built-in buffer against such action. Yet the machine tool slump of the late sixties would heighten only the motives and not the constraints on protectionism at least through 1968. Instead of rising in face of domestic recession, the industry's export dependence fell 76% between 1964 and 1968. And it must be remembered that this exact same period witnessed the second major surge of foreign imports (see Figure 4.6). Between 1964 and 1970, import penetration levels would rise 62% to reach over 13% of the domestic market.

Figure 4.5

Export Dependence* versus Order Backlogs**: All Machine Tools

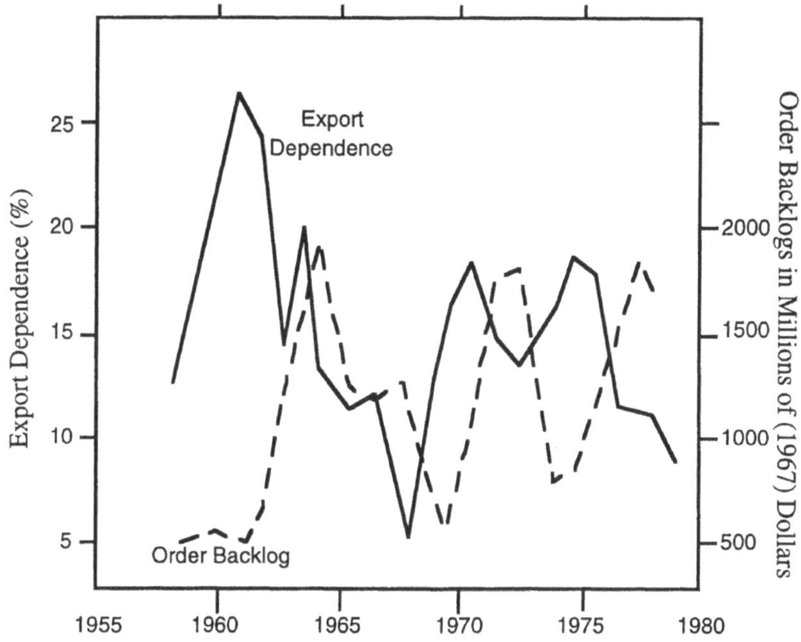

* Manufacturers' Series

** Lagged one year to make direct comparison of one year's orders and their shipment the next. Expressed in constant (1967) dollars.

Sources: Export Dependence – see Figure 4.3
Order Backlogs – NMTBA, "Industry Estimate of Machine Tool New Orders, Cancellations, Shipments, and Backlogs;" U.S. Bureau of Labor Statistics, *Producer Prices & Price Indexes.*

Figure 4.6

Export Dependence versus Import Penetration:
All Machine Tools (Industry Series)

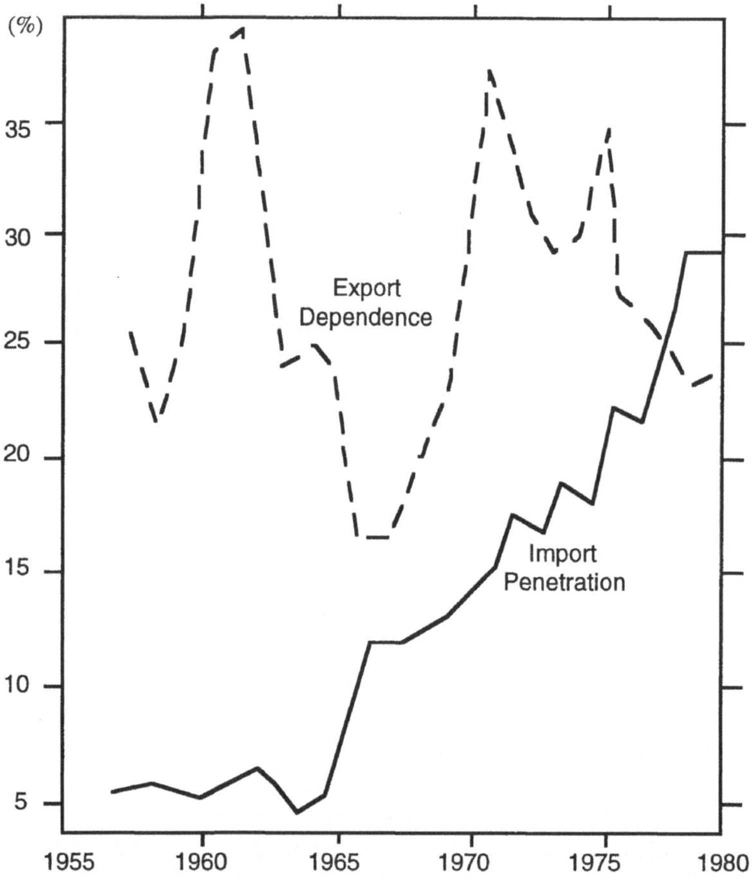

Sources: see Figures 3.1 & 4.3

Hence the late 1960s served as a fertile breeding ground for protectionism. Motives for trade protection — economic recession and import penetration — flourished at the same time that constraints on protectionism — export dependence — waned. The end-product of these conditions was the intensification of protectionist sentiments within the machine tool industry through the sixties that culminated in the 1968 proposal for explicit trade protection in the form of a 15-year import surcharge.

To summarize this part of the argument, machine tool protectionism of the 1950s and 1960s does not contradict the thesis linking export dependence to liberal trade policy. Quite to the contrary, the record substantiates the thesis in that protectionism has been explained, in part, by the decline in export dependence. In particular, substantial deterioration in the industry's dependence upon foreign markets and contingent reductions of constraints against protectionism paved the way for the outbreak of protectionist sentiments in the 1950s and their flourishing in the late 1960s.

The export dependence thesis cannot, of course, be deterministic about protectionist outcomes. Claiming that constraints on protectionism are low is not the same as being able to expect and predict protectionist outcomes. The trade policy views of the machine tool industry in the fifties and sixties were determined, in the first instance, not by levels of export dependence but by other more "immediate causes"; namely, import penetration and recession. Minimal constraints from export dependence provided only the "permissive conditions" that allowed for the ascendance of protectionism.[52]

Yet the purpose of this entire work is to warn against theories of protectionism based solely on immediate causes or, more generally, motivations for protection. Instead, there is a need to examine permissive conditions or, more generally, the structure of constraints on and opportunities for protectionism. Protectionist activity may be constrained in ways not predicted by analyses based solely on immediate causes.

In conclusion, minimal constraints emanating from export dependence opened the door for machine tool protectionism in the fifties and sixties. It remains to be seen whether pressures for protection can be shut out when the constraints of export dependence are high. To demonstrate this latter case, we turn to an explanation of the machine tool industry's adoption of liberal trade policy in the seventies.

1970-1981

To make the case that high levels of export dependence played an important part in the machine tool industry's adoption of liberal trade views, this section first presents statistical evidence attesting to the industry's vigorous export performance in the seventies. It is then established that the industry appreciated the economic importance of these exports. The case culminates with the argument that high levels of export dependence had the political effect of constraining protectionist sentiments.

Export Trends. After deteriorating nearly 20% between 1962 and 1968, U.S. machine tool exports expanded 143% between 1968 and 1975.[53] Over roughly the same period, a number of significant export markets (representing at least $10 million of U.S. exports) experienced even greater expansion. U.S. shipments to West Germany grew 225% (1968-74) and to Australia 244% (1969-75). Exports to Mexico multiplied by four (1968-74) and to Venezuela by six (1968-76). Shipments to Spain and Brazil swelled by factors of ten (1969-75 and 1969-76, respectively).[54] Other major markets for U.S. machine tools included France, Italy, Britain, Canada, and Japan.

Yet one of the most fertile export markets for U.S. tools in the early and mid-1970s was the Soviet Union and Eastern Europe.[55] Prior to 1973, export shipments of virtually all machine tools to socialist countries and other selected countries were restricted by COCOM (Coordinating Committee on Export Controls) and unilateral U.S. national security export controls. Exports of all sophisticated machines — including NC tools — were restricted for all destinations except Canada. As late as 1970, shipments to socialist countries represented less than 3% of U.S. machine tool exports and also less than 3% of all machine tool imports by socialist countries.[56]

Then, as a result of NMTBA pressure, between June 1972 and May 1973 most standard, general purpose machines were removed from control lists. Restrictions remained on NC tools and other sophisticated, specialized machinery, but the partial easing of controls was enough to set off a flurry of export activity. Between 1967 and 1975, U.S. exports to socialist countries multiplied 25 times. These shipments came to represent 18% of all U.S. machine tool exports and over 10% of socialist country machine tool imports. Most of these sales (87% in 1975) were made to the Soviet Union. Indeed in 1974 and 1975, the Soviet Union was the number one export market for the U.S. in machine tools compared to being number four in 1973, number eight in 1971, and number 14 in 1967.[57]

As a result of these export successes, the industry's dependence upon foreign markets heightened. After falling 76% between 1964 and 1968, the industry's export dependence jumped 112% from 1968 to

1971 (manufacturers' series). In 1971 over 35% of the industry's shipments went abroad — a far cry from 1968's levels of 16% in the industry series and 5% in the manufacturers' series. Through the 1970s, the industry shipped an average of 29% of its machine tools to foreign markets. (As discussed earlier, the fluctuations in export dependence through the seventies were largely counter cyclical effects of the machine tool business cycle.) And the larger machine tool builders led the U.S. export binge. In the early seventies, Cross exported 34% of its production.[58] In 1975, Acme-Cleveland and Giddings & Lewis shipped 22% and 23% of their shipments abroad, respectively.[59] And Cincinnati Milacron garnered 44% and 40% of its sales from foreign markets in 1971 and 1975, respectively.[60]

Reliance Upon Exports. Since 1971, the machine tool industry has gone out of its way to stress the importance of exports to the business. Exports have been repeatedly recognized as an important source for machine tool employment, output, and funds for R&D and capital expansion.[61] More importantly, exports have been considered vital to the very health of the industry. Exports have, in fact, been cited as representing the "difference between bankruptcy and survival" during the machine tool depressions of 1921, 1923, 1938, and 1946-49.[62] Yet it would not be until the 1970s that the industry would come to appreciate the enduring value of exports.

The 1969-1971 machine tool depression was instrumental in heightening the importance of exporting. As expressed by NMTBA Executive Vice President James A. Gray:

> The companies that were involved in the export market in the last depression had far less trouble than those that were not. In order to sustain us in the next low point in cycle, we've got to anticipate it now and have a foreign market that we can turn to keep our factories busy.[63]

The importance of foreign markets was magnified by the recognition that overseas demand ran counter cyclical to domestic machine tool demand.[64] While U.S. machine tool shipments dropped 45% between 1967 and 1971 (from $1.8 billion to $1 billion in current dollars), foreign consumption for machine tools increased over 40% over the same period.[65]

Indeed, by the early 1970s a fundamental transformation had taken place regarding the position of the U.S. machine tool market vis-à-vis the world market. In 1966-67, the U.S. held a commanding position in world machine tool consumption — domestic users accounted for nearly 30% of machine tool consumption world-wide (see Figure 4.7). If there was ever a time when the industry could afford to insulate itself — to turn away from foreign markets, turn out foreign imports, and rely solely on a vast and protected domestic market — the late 1960s was the time. Yet in the early 1970s, the relative importance of the U.S. market significantly deteriorated. By 1971, the U.S. market accounted for only 10% of world machine tool consumption.

Figure 4.7

U.S. Machine Tool Consumption as a Percentage of World Machine Tool Consumption

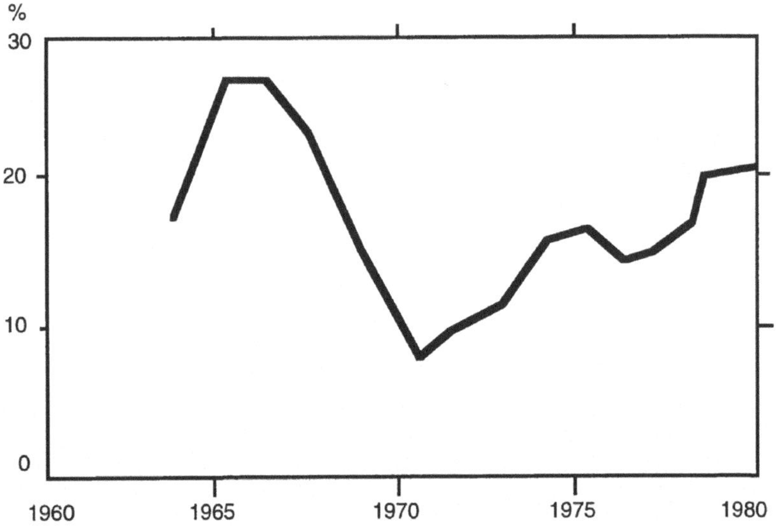

Source: American Machinist

This relative "shrinking" of the U.S. market has not been lost on industry officials. Through the 1970s and into the 1980s, Frank J. Austin, Giddings & Lewis Executive Vice President and Chief Operating Officer, among other machine tool officials, acknowledged that "while the U.S. machine tool market has been oscillating with very little real growth since the mid 1960s, the world market has grown steadily."[66] Figure 4.8 depicts the stunning secular growth in world machine tool consumption. The message, as repeatedly expressed by NMTBA President Gray, for the industry has been clear: "the only way to maintain a viable machine tool industry is to become so closely tied to the world market that we gain by its steady upward growth."[67]

Figure 4.8

World Machine Tool Consumption

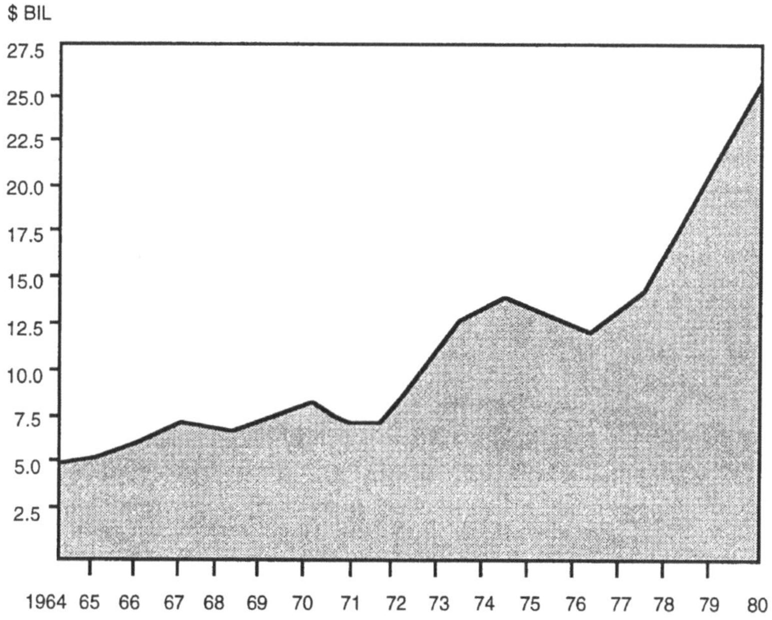

Source: American Machinist and Uniform International Statistics on Machine Tools

This argument is neatly summarized in a line of thought expressed time and time again by the NMTBA:

> . . . the world machine tool market has been growing steadily and is almost unaffected by the ups and downs of individual national machine tool demand cycles. It is simply the result of the counter-cyclical effect of combined markets. The world machine tool

> market is the answer to our industry's problems. If the American machine tool industry would really market its products on a global scale, it could become almost immune to the vagaries of the domestic business cycle. As a result, our industry could better absorb sudden domestic demand peaks and meet our customers' immediate machine tool needs. *All this brings us to the inescapable conclusion that the health of the American machine tool industry is heavily conditioned upon our industry's ability to expand its global markets.*[68]

Constraints on Protectionism. Having established that the machine tool industry came to appreciate the strategic value of exports, the final task in making this argument is to establish whether and how export dependence worked to constrain protectionist pressures within the industry in the 1970s. The contradiction between import protection and export reliance has, in fact, been recognized by the industry since the 1950s. Even when the industry considered and advocated protection, it was always cognizant of the potential costs to be incurred to its exports. The 1955 NMTBA "National Policy" statement explicitly acknowledged the problem:

> Machine tool imports should be held to a minimum. Meanwhile, however, a substantial export market for machine tools would mean larger United States capacity for use in times of emergency. The problem is therefore to maintain the industry's traditional export market while at the same time restricting the importation of foreign imports to a minimum. It will at once be apparent that *there is an inherent inconsistency between restrictions on machine tool imports and free machine tool exports.* The machine tool industry recognizes this inconsistency and understands that *compromises must be made.*[69]

When tariff increases were considered in 1958 and 1959, the industry was split between those advocating the increase and those that did not for fear of "retaliation against the export business."[70]

Even the 1968 import surcharge system was proposed in a context where the tradeoffs of protection were very much considered. Again, the industry was very concerned about the implications for its export markets. As stated by Brown & Sharpe First Vice President and NMTBA First Vice President Henry Sharpe before the 1968 congressional panel:

> the question is: How do you protect this American industry without so upsetting our trading partners and their own machine tool industries that it causes needless ruffling of the water.[71]

Given these concerns, the industry opposed the Mills Bill quota legislation which set rigid quantitative limits on imports. In the industry's eyes, its surcharge system was "less likely to invite retaliation" since the system allowed for the growth of imports but only, of course, at progressively higher cost (i.e., staged tariff increases).[72] In the industry's words, the surcharge proposal "provides a cushion and avoids doing something violent and quick that would cause needless antipathies."[73]

Hence even when the machine tool industry went down the protectionist road in the fifties and sixties, it did so with an eye over its shoulders on its exports. While the concern for exports did not, at this point, induce the industry to forsake its protectionist positions, it is significant that this concern *did* work to moderate its trade policy views. Constraints on protectionism from export dependence were, of course, deteriorating during this period. When export dependence levels substantially increased in the 1970s, the industry's cautious journey toward protectionism would be stopped dead in its tracks and the industry would begin backtracking toward liberal trade positions.

The industry's abandoning of its import surcharge system can be directly linked to the heightening of export dependence in 1970. At the 1970 congressional hearings where the NMTBA backed away from its proposal of two years prior, Sharpe was clearly mindful of the costs of protection:

> we still must be extremely careful to do whatever we do with an eye to not causing unnecessary antipathies among our trading partners . . . . If a retaliatory, cool attitude toward the United States is fostered because of the things that we do here, this will work to our detriment.[74]

Such costs of protection have, as we've seen, been recognized since the 1950s. Yet in 1970 these costs had a deterring effect on protectionist sentiments because the industry had developed a keen sense that it had become closely tied to the world economy and that it could no longer risk a protectionist position. Sharpe rejected protection against foreign suppliers because "we, as an international industry, are very sensitive to the desirability of participating in those foreign country's own markets."[75] Regarding a query as to the reasons for the industry's adoption of a "nice guy" trade position, Sharpe responded: "The reason I want to be is because the machine tool business is an international business, and the greatest benefits come from the world-wide application of a given [machine tool] design in as many different markets as can participate actively in it."[76]

The basis for the policy change of 1970 was made explicit: A protectionist position would entail costs to exporting and the industry was so dependent upon such external links that such costs were unacceptable. In the late 1960s when the U.S. market accounted for one-third of world machine tool consumption and when less than 5% of U.S. machine tool production was sent abroad, the industry might have been in a position to absorb the repercussions of a protectionist policy. When in 1970 with the U.S. consuming only 10% of the world's machine tools and with 28% of its production destined for foreign markets, the luxury of a protected market could no longer be afforded.

From 1970, the industry's reliance on foreign markets was the underpinning for its maintenance of liberal commercial policy positions. The industry established its strong advocacy of free trade in hearings on the 1974 Trade Act where Cincinnati Milacron President and NMTBA First Vice President Philip Grier reported:

> NMTBA is convinced that a return to protectionist trade policies would result in retaliation by our trading partners and ultimately *deprive* the U.S. of appropriate participation in *expanding world markets*.[77]

During the Tokyo Round of negotiations, industry officials observed:

> Measures advocated by protectionist forces in this country might spell ruination of the machine tool industry because we will be *unable to competitively market our products overseas.* (NMTBA President James Gray before congressional hearings on Export-Import Bank Authorization, April 7, 1978, p. 281)

> We can fight this foreign competition . . . [But] we must counter its drive with positive measures of our own rather than a negative position of *undercutting exports* . . . we must battle on a positive rather than a negative note." (Leon B. Musser, Vice President for marketing and corporate development, Kearney & Trecker, cited in *Iron Age*, August 27, 1979, p. 84.)

In 1980 hearings on U.S. trade relations with Japan, Giddings & Lewis Executive Vice President Frank Austin testified:

> Although we certainly sympathize with the objective of keeping as many American workers employed as possible, we must disagree with restrictive import quotas as a method of achieving the goal. We have seen the *unfortunate consequences of restrictive trade practices* in the past and do not offer them as a solution to the current United States-Japanese trade imbalance.[78]

To conclude, there is strong evidence that concern for the ability to market its products abroad played an important if not decisive role in the machine tool industry's advocacy of relatively liberal commercial policy positions in the 1970s. It has been shown that the industry came to rely heavily on foreign markets for its sales beginning in 1970. It was also established that the industry came to recognize that such foreign sales were vital to its economic health. Finally, evidence has been presented attesting to the industry's recognition that protection would entail costs for its exports, and that given its heightened economic vulnerability and dependence such costs ruled out protectionist indulgence.

Some of the statements quoted above particularly cite foreign retaliation as a potential cost of protection while others refer to unspecified negative ramifications for exports. It may be too much to expect reference to the specific price, supply, and other costs of protection discussed in Chapter I. Such price and supply mechanisms, as induced by foreign trade, have, however, been recognized by U.S. machine tool builders. For instance, as will be discussed later, the competitive prices of foreign machine tools forced domestic builders to cut their own prices. A strategy designed to combat foreign products at home also produced new price-competitive machines with great marketing potential abroad. Remove foreign price competition by trade protection and such price-consciousness would weaken as would prospects for international marketing.

Yet the important point of this discussion is that the constraints of export dependence, whether general or specific, made a substantial impact on machine tool trade policy in directions postulated. Low levels of constraint emanating from export dependence posed little obstacle for machine tool protectionism in the fifties and sixties, and high levels imposed critical trade policy constraints that helped move the industry toward liberal commercial policy positions in the seventies.

ALTERNATIVES TO PROTECTION

Having outlined a critical reason for not pursuing trade protection, it would be appropriate to briefly discuss the policy alternatives to protectionism that the machine tool industry pursued in responding to intense foreign competition. Such non-protectionist policy options do not emanate directly from conditions of high export dependence any more than protectionism does from low export dependence levels. Yet the machine tool industry's responses to international competition, largely undertaken beginning in the mid-seventies, help to define some of the policy alternatives to protection for export dependent industries generally. These alternatives entail initiatives in the realms of both *public* policy and *business* policy.

Public Policy

While the machine tool industry did not go to the government pleading for trade protection in the 1970s, it did not shy away from lobbying for government legislation considered to serve its interests. In virtually every case save one (product liability), NMTBA public policy concerns were explicitly related to the enhancement of the industry's global competitiveness. A significant indicator of the value that the industry placed upon exports is the fact that most NMTBA government policy concerns fell into two major categories: trade liberalization and export promotion.

Trade Liberalization. The industry sought to free its exports of restrictions through the liberalization of foreign trade barriers, the extension of most favored nation (MFN) status, and the reform of U.S. export controls. Whether advocating protectionist or liberal trade positions, the NMTBA always called for the reduction of foreign tariffs. Beginning in 1968, the industry also called for the elimination of foreign nontariff trade barriers.[79]

Machine tools have, in fact, been centrally involved in U.S. tariff negotiating efforts, particularly with Japan. In the major trade agreement diffusing the U.S.-Japan trade crisis of 1971-72, Japan agreed to reduce duties on machine tools and 30 other industrial products. Machine tool tariffs were reduced from a 7.5%-15% range to 6.5-13.5%.[80] In response to U.S. pressure to reduce the U.S.-Japan trade gap of 1977-78, Japan moved to reduce its machine tool tariffs by 14% prior to negotiated reductions in the Tokyo Round.[81] As part of the Strauss-Ushiba agreement that unshackled the Tokyo Round negotiations, Japan agreed to duty reductions on more than 300 products including a 25% cut for machine tools.[82]

In 1979, the United State Trade Representative (USTR) cited the U.S. machine tool industry as being among a dozen industries which would experience "the most significant increases in business" as a result of the Tokyo Round trade pact. The USTR predicted that metalworking equipment exports would be boosted by $61.2 million per year while imports would increase only $49.8 million. The agreement was to result in a net gain of 1097 job opportunities for the industry. Machine tools were to particularly benefit from the liberalization of the Canadian machinery program and the phasing out of many "Made in Canada" tariff lines.[83]

Following the conclusion of the Tokyo Round, the U.S. machine tool industry continued to be a beneficiary of U.S. pressure to liberalize the domestic Japanese market. In February 1982, MITI moved to lower machine tool tariffs to at least a 3% ad valorem rate.[84] Then just prior to the June summit of Western leaders in Versailles, the Japanese included machine tools among the 94 products slated for tariff elimination.[85]

The NMTBA sought to expand foreign markets not only through the reduction of foreign trade barriers but also through the extension of MFN status to socialist countries. Beginning in 1968, the industry supported legislation to expand East-West trade.[86] The NMTBA vigorously supported the extension of MFN status to Czechoslovakia, Hungary, Rumania, and China.[87] Industry efforts to further expand its trade with socialist countries were, however, hindered by U.S. export controls.

Indeed, as much NMTBA lobbying energy has been devoted to the reduction of *U.S.* export controls as to the reduction of foreign trade barriers. Beginning in 1958 and with increasing intensity, the industry sought to reform the "national security" and "foreign policy" export controls imposed on machine tool shipments to socialist countries and selected other countries under allied (COCOM) and unilateral (Export Administration Act) regulations. In light of alleged "lost sales" to European competitors and/or violators, the NMTBA demanded the uniform enforcement of COCOM controls. The association also lobbied for simplified, consistent, and reliable licensing procedures to which restricted products were subjected. Finally, the industry sought the total elimination of controls on products that were available to socialist countries either internally or from non-COCOM suppliers.[88]

The association reserved, however, its greatest hostility for the export restrictions encompassed in the Jackson-Vanik amendment (Title IV) to the Trade Act of 1974. From the amendment's initial proposal, the NMTBA vehemently opposed and sought the unqualified repeal of the linkage between the provision of export credits and Soviet emigration policy.[89]

Export Promotion. Besides the liberalization of machine tool exports from domestic as well as foreign restrictions, the industry sought government aid to promote exports. Central to these latter efforts was the industry's support, beginning in 1956, of export financing by the Export-Import Bank. Besides the continuation of the Bank's authority, the NMTBA lobbied for increased Export-Import resources to provide export credits competitive with those of foreign competitors, and for the disencumbrance of Export-Import financing to socialist countries.[90]

The NMTBA also lobbied for governmental export assistance other than through the Export-Import Bank. Beginning in 1968, the association heartily endorsed the export promotion activities of the Commerce Department which included the provision of international marketing advice and the sponsorship of trade missions and exhibits.[91] Also in 1968, the industry recommended that export income be exempted from Federal income tax. When legislation along these lines was proposed and enacted in 1971 in the form of the Domestic International Sales Corporation (DISC), the NMTBA supported and subsequently defended it.[92]

Beginning in 1979 the industry also vigorously lobbied for export trading company legislation. The NMTBA championed legislation amending U.S. antitrust laws to permit U.S. corporations to combine their efforts for export purposes.[93]

Beyond the machine tool industry's efforts regarding trade liberalization and export promotion, the NMTBA lobbied for more "domestic" pieces of legislation that were, nevertheless, expected to contribute to an effective response to foreign competition. Such public policy issues included tax breaks for R&D expenditures, government support for technical training, and the liberalization of depreciation write-offs and investment tax credits.[94] Among the arguments for each proposal was the stimulus provided for exports. For instance, among the reasons the NMTBA backed the Reagan administration economic program and especially its accelerated depreciation provisions was for its contribution to "making America once again fully competitive in world markets."[95]

In summary, the machine tool industry sought a wide range of government cooperation and aid in lieu of trade protection. In the industry's own words: "The American machine tool industry is not looking for special favors or preferential treatment. All we want is freedom to compete equally with the other nations of the world."[96]

Business Policy

Despite NMTBA efforts to shape public policy to serve its interests, the industry fully understood that it must be "prepared to devote its own resources to the development and maintenance of international markets everywhere in the world."[97] By reviewing the independent initiatives of individual machine tool firms, a diverse set of general business strategies can be construed as representing alternatives to trade protection. These strategies, which largely arose beginning with the mid-seventies, included: export expansion, product development, cross-national ties, capital expansion, technological development, industrial concentration, and horizontal integration.

Export Expansion. Parallel to efforts to gain official government assistance for exporting, the industry undertook its own initiatives to promote exports. The international trade activities of the NMTBA highlight these efforts. The association held seminars and workshops to train firms in international financing, export licensing, and contract negotiating; conducts market research to locate new and promising export markets; organizes trade missions to gain footholds in these markets; sponsors foreign exhibitions to display U.S. machine tools overseas; organizes reverse trade missions to bring foreign buyers to the states; and attracts foreign visitors to its biennial International Machine Tool Show in Chicago. The NMTBA has a full-time staff of three to promote exports abroad.[98]

Product Development. As described in Chapter III, the Japanese success in the U.S. market was based, in part, upon the exploitation of a market "gap" in the product lines offered by U.S. builders. In particular, Japanese suppliers took advantage of their U.S. counterparts' weakness in medium range, cost-efficient, highly versatile machine tools. In direct response to this development, U.S. builders began to take a second look at its machine tool designs and committed substantial engineering resources towards reducing unit costs while maintaining versatility. These efforts first came to fruition in the late 1970s and five developments stand out.

The first major domestic machine designed to counter foreign competition was Kearney & Trecker's April 1978 introduction of its Milwaukee-Matic 180 horizontal NC machining center. The small-sized MM-180 was designed around a cost target of $125,000. While offering more features, the machining center undercuts its Japanese competition by $10,000. The MM-180 took two and a half years to develop and test, and instrumental to its designing was a visit by Kearney & Trecker's engineering team to Japan in May 1976.[99]

In early 1980, Cincinnati Milacron introduced a new line of small, low-cost NC lathes sized and priced to compete directly with some of Japan's largest and most innovative lathe builders. In a sincere parody of the 1980 Democratic primary contest between Carter and Kennedy, a Milacron official remarked, "We are not unaware of the fact that we will be going head-to-head with the Japanese . . . . We're not afraid of them. We can whip their asses."[100]

When the Japanese penetrated the NC lathe market in 1978, Pratt & Whitney Machine Tool Division (of Colt Industries) was marketing an eight-tool lathe introduced in 1972 and priced well above $100,000. In contrast, the Japanese product was priced around $95,000 and had 12 tools, more sophisticated controls and higher spindle speeds. In the words of a Pratt & Whitney sales manager, "we were trying to sell a machine with far less for more money." In response, in 1981 Pratt & Whitney developed a new version of its small lathe. Designed to "undercut" the Japanese, the new lathe offers a 12-tool capacity at $85,000.[101]

Having already introduced its new lathes in 1980, Cincinnati Milacron unveiled a medium-sized NC horizontal machining center in

September 1981. The result of four years of development, the relatively sophisticated yet price-competitive ($215,000) machine was conceived to combat the machining centers of Yamazaki Machinery Works.[102]

Finally, in early 1982 Warner & Swasey introduced its low-cost "world lathe" which is "viewed in the industry as an attempt to halt and then reverse Japanese inroads into Warner & Swasey's bread-and-butter business: NC turning." Making its "world lathe" all the more noteworthy was Warner & Swasey's selection of General Numeric NC controls as standard equipment. Controls from the U.S. joint venture of Fujitsu Fanuc and Siemens were chosen over those of Allen-Bradley and General Electric, Warner & Swasey's traditional suppliers.[103]

The opposite of product development is the phasing out of product lines. In face of acknowledged comparative advantage abroad, U.S. builders dropped production and conceded markets for many conventional machines such as engine lathes, grinders, shapers and planers.[104]

Cross-National Linkages. To further product development and product discontinuance, among other things, U.S. machine tool builders increasingly entered into unprecedented marketing, licensing, and joint venture agreements with foreign builders. U.S. builders have had longstanding cross-national agreements, but they were designed to *export* U.S. goods or license U.S. technology abroad. In contrast, beginning in the mid-seventies U.S. builders entered into agreements to *import* foreign machine tools and foreign machine tool technology. Table 4.5 notes some of the major marketing arrangements that were entered into. In general, these agreements broadened the product offerings of the U.S. builders. For instance, the Toshiba boring machines did not compete with other Kearney & Trecker tools[105] and the Okamoto gear-grinders supplemented Barber-Colman's gear-hobbing and gear-shaping lines.[106]

Table 4.5

Cross-National Marketing Arrangements

Year	*U.S. Importer*	*Foreign Supplier*	*Product*
1976	Pratt & Whitney	Fujitsu Fanuc) (Japan	low-cost machine tools
1977	Kearney & Trecker*	Toshiba Machine (Japan)	large boring machines
1977	Barber-Colman	Okamoto Machine Tool Works (Japan)	gear-grinders
1978	Pratt & Whitney	Mitsui Seiki (Japan)	medium-sized machine tools
1978	Pratt & Whitney	Fujitsu Fanuc	vertical machining centers
1978	Kearney & Trecker*	Daewoo Heavy Industries (South Korea)	vertical milling machines
1980	Pratt & Whitney	Hamai (Japan)	machining centers
1980	Barber-Colman	Toshiba Machine	hobbing machines
1982	Landis Tool**	AB Bofors-Uva (Sweden)	grinding machines

(continued)

Table 4.5 (continued)

* Subsidiary of Cross & Trecker
** Subsidiary of Litton

Sources: *American Machinist*, April 1977, p.33; October 1977, p.47; April 1980, p.65; December 1981, p.55; *American Metal Market/Metalworking News*, April 14, 1980, p.77; March 8, 1982, p.19; *Iron Age*, August 21, 1978, p.14; September 11, 1978, pp. 16,33; *Japan Economic Journal*, February 15, 1977, p.7; April 15, 1980, p.7.

In many cases, marketing agreements were made for U.S. builders to take advantage of low-cost foreign machine tools. The Pratt & Whitney ties to Fujitsu Fanuc, Mitsui Seiki, and Hamai all of Japan fall into this category. Pratt & Whitney's President, Irv H. Pennan, acknowledged that there was no way his establishment could build the marketed machines at competitive prices in the U. S. For instance, the Fanuc vertical machining center offered three-axis with 15 tools at a phenomenal price of $79,000.[107]

In the Kearney & Trecker-Daewoo case, the marketing arrangement was designed to supplant a U.S. product line. In recognition of lower production costs abroad and in order to concentrate on higher technology products, Kearney & Trecker sold its licenses for conventional milling machines to the South Korean firm. As Daewoo went into production, Kearney & Trecker phased-out its production of the machines. The Daewoo machines then were marketed in the U.S. and elsewhere. The agreement evinces efforts among U.S. builders to shift energies towards the more sophisticated end of the machine tool market.[108]

Table 4.6 indicates the major cross-national licensing agreements and joint ventures designed to import not foreign machine tool products but foreign machine tool technology. Some of the agreements were struck for U.S. builders to gain cost-effective foreign technology (e.g., Olofsson-Kanzaki Kokyokoki). Yet others involved the importation of advanced, state-of-the-art technology. It was noted earlier that the U.S. had fallen behind both Europe and Japan in unmanned machining. An ameliorating step was taken in this regard by Warner & Swasey's 1979 agreement to license and market turning centers from Emag Maschinenfabrik. These machines, equipped with twin spindles and handling arms, were designed to operate untended for long production runs.[109]

Table 4.6

Cross-national Licensing Agreements and Joint Ventures (U.S.-Based)

Year	*U.S. Importer*	*Foreign Supplier*	*Product*
1978	Morey Machinery	Graffenstaden Division, CIT Alcatel (France)	tracer lathes
1978	Olofsson	Kanzaki Kokyukoki Mfg (Japan)	low-cost head changer
1978	Warner & Swasey*	EMAG Maschinenfabrik (West Germany)	NC lathes
1979	Monarch	Morando (Italy)	vertical turning machines
1979	Warner & * Swasey	EMAG Maschinenfabrik	unmanned turning machines
1980§	Strassman Machinery	Ikegai Iron Works (Japan)	NC lathes
1980§	Inova	Sodick (Japan)	electrical discharge machining
1980§	John Brown	Taiyo Seiki (Japan)	turning machines

(continued)

Table 4.6 (continued)

Year	*U.S. Importer*	*Foreign Supplier*	*Product*
1981	Jones & Lamson**	J.G. Weisser Sohne (West Germany)	twin spindle NC lathes
1981	Ex-Cell-O	Stama Maschinenfabrik	vertical machining centers
1982§	Warner & Swasey*	Murata Machinery	machine tools

§ U.S.-based joint venture
* Subsidiary of Bendix
** Subsidiary of Textron

Sources: *American Machinist*, July 1978, p.35; September 1980, p.61; November 1981, p.63; December 1981, p.63; December 1981, p.55; *American Metal Market/ Metalworking News*, June 30, 1980, p.32; November 16, 1981, p.12; December 7, 1981, p.12; Iron Age, September 4, 1978, p.15; December 4, 1978, p.19; August 27, 1979; January 15, 1979; December 17, 1979; September 8, 1980, p.15; September 15, 1980.

Table 4.7

Capital Expansion Ratios* for Leading Machine Tool Builders (%)

	1975	1976	1977	1978	1979	1980	1981
Acme-Cleveland	3.5	4.4	2.1	3.1	4.3	6.4	4.1
Bendix/Industrial			4.9	3.8	3.4	4.0	
Warner & Swasey	2.5	1.5	4.2	3.3			
Cincinnati Milacron	2.7	4.2	4.3	3.9	5.2	5.4	5.3
Cross & Trecker	1.6	3.1	2.1	3.0	3.1	5.3	6.8
Ex-Cell-O	4.2	5.1	7.0	3.9	4.4	3.8	3.7
Giddings & Lewis	3.6	2.1	3.7	5.4	4.1	4.2	
Litton/Industrial Systems	2.3	2.5	4.9	8.3	7.8	8.7	11.5
Textron/Machine Tool					4.0	4.7	5.5

* ratio of expenditures for plant and equipment to total sales

Sources: *Annual Reports*

Capital Expansion. In the late seventies, the machine tool industry orchestrated an extraordinary increase in capital expansion. Between 1976 and 1980, the industry's expenditures for new plant and equipment grew at an average annual rate of 38% (see Figure 4.9). Virtually every and all major machine tool firms embarked on ambitious broad-based expansion programs including Cross & Trecker, Ex-Cell-O, Ingersoll Milling Machine, Cincinnati Milacron, Jones & Lamson (Textron), and LaSalle (Acme-Cleveland).[110] Table 4.7 largely confirms this growing commitment among leading machine tool builders to invest greater proportions of sales for capital expansion. In a 1980 NMTBA survey, 85% of responding members indicated they had begun expansion programs during the past 18 months.[111]

Capital spending by the machine tool industry as a whole not only outpaced that of other manufacturing industries (see Figure 4.9), but represented a significant break from the industry's past performance. Past heavy investment periods in the mid-1950s and mid-sixties were only to be met by machine tool recessions that resulted in idle capacity. The industry subsequently became very wary of capital expansion. Expenditures for new plant and equipment grew only at an average of 13% per year between 1966 and 1975[112] — the very time when foreign imports begin making their serious inroads.

The reason for the dramatic change in the late seventies was intimately related to pressures from international competition. In the short-term, capacity expansion was aimed at reducing backlogs to combat foreign imports. Inadequate expansion between 1966 and 1975 resulted in the building-up of order backlogs that invited the penetration of foreign suppliers. In the longer term, the expansion was motivated by the strength of the domestic market and efforts to gain an edge in international competition.[113]

Figure 4.9

Average Annual Capital Investment Growth, 1976-1980

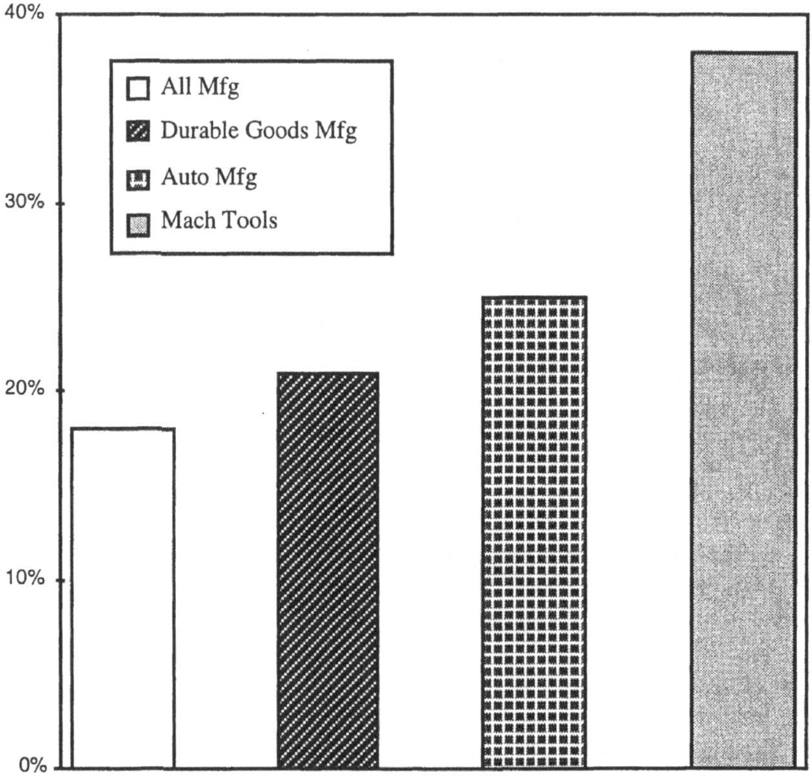

Source: National Machine Tool Builders' Association, U.S. Bureau of Economic Analysis

Technology Development. Beginning in the late seventies, the industry also took great strides on the technology front. While by no means approaching the performance of "high-technology" firms, virtually all major machine tool builders committed unprecedented resources to R&D — including Cincinnati Milacron, Ingersoll, Jones & Lamson, Kearney & Trecker, Giddings & Lewis, Cross, Bendix, LaSalle, and F. Jos. Lamb.[114] Among the most notable efforts were on the part of Cincinnati Milacron. By spending over $100 million on R&D between 1977 and 1981, over 40% of Milacron's sales consisted of products manufactured only in the past five years.[115] Cincinnati Milacron increased its R&D expenditures over 300% between 1975 and 1981, and in 1981 announced the construction of a $6.8 million R&D center.[116]

These expenditures were aimed not only at product innovation but also at advances in process technology. For years such Japanese machine tool giants as Fujitsi Fanuc and SNK utilized such advanced manufacturing processes as unmanned machining, automated storage and retrieval, and computer-aided-design (CAD).[117] To bolster its international competitiveness, U.S. builders also moved to gain greater manufacturing capability and productivity through advanced process technology. In 1981, Cincinnati Milacron installed an automated storage and retrieval system and 37 CAD terminals. CAD is used to reduce a machine's parts, assembly time, and cost. Twenty-seven more terminals were scheduled for 1982 as well as the installation of automated machining cells and FMS.[118] Bendix has used CAD, initially as a drafting tool, since 1974. Bendix systems not only helped to design machine tools but also generated programs to direct the machines — an important step towards computer-aided-manufacturing (CAM).[119]

CAD systems have advanced from drafting tools and the design of electrical schematics to analyzing the structural design of machine tools. A leader in this advanced capability was F. Jos. Lamb. Lamb's Structural Kinematics Division utilized CAD to reduce deformation and vibration of machine tool parts and increase machine durability and accuracy.[120]

Industrial Concentration. In 1966, Henry J. Sang of SRI predicted that by 1975 twenty large companies would make up the machine tool industry. In fact, in 1980 there were an estimated 500 machine tool firms — a slight increase over the early 1970s.[121] Nevertheless, machine tool mergers and acquisitions overtook the industry in the late seventies. In 1977 White Consolidated Industries purchased the machine tool operations of Sundstrand to create the twelfth largest U.S. machine tool concern.[122] Nineteen seventy-seven and 1978 also saw Bendix attempt to acquire Ex-Cell-O and an unsuccessful merger between Giddings & Lewis and Motch & Merryweather (Cleveland, Ohio).[123]

The concentration of the industry quickened beginning in 1979. Among the most significant moves was the merger between Hartford Special Machinery and New England Machine and Tool. Gleason Works and Alliance Tool (both of Rochester, New York) merged to become number eleven of large machine tool corporations. Giddings & Lewis purchased Snyder (Detroit) and became number six. Kearney & Trecker combined with Cross to become the third largest machine tool builder. (Cross had already acquired Avery Machine Tool.) And in 1980, Bendix acquired Warner & Swasey to become the second largest producer behind Cincinnati Milacron. (Bendix had previously bought several machine tool concerns under one roof.)

The basis for these mergers and acquisitions was, in general, the need to establish a broad financial base to meet the challenges of the business — challenges which in so many ways were created by foreign competition. In particular, three concerns motivated the concentration. First, companies with broad product lines such as Milacron and Acme-Cleveland were better able to absorb cyclicity in any given product line than more specialized firms. The need to diversify certainly motivated the mergers between Hartford (forming machines) and New England (sophisticated milling), and between Kearney & Trecker (machining centers and manufacturing systems) and Cross (transfer lines). Second, in the gradual moving away from building on small, specialized orders, machine tool firms sought greater capacity and resources to make large quantity runs in order to take advantage of the savings accrued by manufacturing on a larger scale.

Third, the intensity of foreign competition heightened the value of technological sophistication and innovation. Among the stated objectives of the Kearney & Trecker-Cross merger was the combining of Cross' experience in big, complex manufacturing systems (transfer lines) and Kearney & Trecker's background in FMS in order to develop the "manufacturing technology of the eighties" — flexible, medium-to-high volume manufacturing systems. Specialized, technologically sophisticated firms were prime targets of larger companies, For instance, Giddings & Lewis acquired Snyder's high-tech capabilities to move into the FMS field. More generally, the pressure for greater technological advancement required R&D commitments that only the larger firms can afford.[124]

Horizontal Integration.[125] The quest to bridge soft spots in any given product line led machine tool builders to diversify their machine tool product lines, sometimes through industrial concentration. This same quest also led many builders to diversify outside of the machine tool business. Many firms came to produce such machine tool "peripherals" as machine tool electronics, cutting tools, cutting fluids, and grinding wheels. In-house NC control production were established by Bendix, Bridgeport (Textron), Cincinnati Milacron, Giddings &

Lewis, Kearney & Trecker, Warner & Swasey, and White-Sundstrand.[126] Cincinnati Milacron was a major force in industrial robotics, and Pratt & Whitney and Bridgeport moved into the robot business as well.[127]

The emphasis on FMS intensified the trend toward horizontal integration since such systems involve much more than just machine tools. Firms had to quickly pick up expertise in computer interfacing, automatic inspection, and workpiece handling. Indeed, many machine tool firms moved towards providing "turnkey" manufacturing systems that offer not only machining, inspection and workpiece transfer, but also assembly, storage and inventory.[128]

Major machine tool firms also moved into areas completely unrelated to machine tools. The most prominent example of this trend was Cincinnati Milacron which produced plastics machinery, semiconductor materials, and chemicals. (Milacron sold its chemical operation in May 1980.) The percentage of Cincinnati Milacron sales represented by machine tools diminished from 72% in 1970 to 64% in 1979. Ex-Cell-O's industrial equipment sales also declined as a proportion of total sales; from 59% in 1973 to 34% in 1981. And the percentage of Giddings & Lewis sales consisting of machine tools dropped from 96% in 1975 to 68% in 1980.[129]

In summary, two broad objectives can be construed from the diverse business strategies described above. In seeking to gain an edge over foreign competition, U.S. builders sought to enhance their *technological sophistication* through product development, cross-national linkages, capital expenditures, industrial concentration, and, of course, technology development. Product development, cross-national linkages, industrial concentration, and horizontal integration also served the *diversification* of the industry. Flexibility and broad-based viability was sought through diversification in products and markets served. As described by the industry, these business strategies along with the industry's public policy efforts — and not trade protection — served as "the most effective means of meeting the increasing challenge of imports in our domestic market and of increasing our own export sales efforts."[130]

The 1980s

An important attribute of this analysis is its dynamic character; when the independent variable changes, so too does the dependent variable. Just as the constraints of export dependence heightened in the seventies, they can subsequently weaken. Such a deterioration, in fact, took place in the early 1980s, and the remainder of this chapter is devoted to analyzing its repercussions.

In aggregate terms, the machine tool industry's export performance suffered terribly beginning in the mid-late 1970s. Figures 4.3 and 4.4 exhibit the weakening that took place in the industry's dependence upon exports. The manufacturers' series reveals that export dependence reached its lowest levels since 1968 — the year the industry took its greatest stride toward trade protection. Indeed, for the crucial metal-cutting segment of the industry, export dependence levels fell below those for 1968.

The export performance of the leading machine tool builders suffered along with the rest of the industry. The proportion of Cross shipments exported declined from 34% in the early seventies to 10% in 1979. Acme-Cleveland exports dropped from 22% of shipments in 1975 to 6% in 1979. Over the same period, the ratio of exports to shipments for Giddings & Lewis shrank from 23% to 5%.[131]

The reasons for this export decline are numerous and complex ranging from exchange rate appreciation and foreign trade barriers to domestic inflation and the erosion of international competitiveness. Two of the most prominent problems that have seriously hindered U.S. machine tool exports concern government (both foreign and domestic) policy: export financing and export controls.

With price tags ranging from $50,000 to several millions of dollars, machine tool purchases constitute major industrial investments. As such, machine tool sales at home or abroad are very sensitive to terms of credit. Beginning in the late seventies, U.S. builders have repeatedly complained about the uncompetitiveness of U.S. export financing and particularly Export-Import Bank export credit (see discussion above). These complaints reflected the generally disadvantageous position the U.S. held in the international "terms war." In a 1977 NMTBA survey, 18 of 55 machine tool firms reported lost foreign orders due to uncompetitive financing. The lost orders amounted to $251.4 million in sales or equivalent to 56% of U.S. machine tool exports for 1977.[132] The industry abounded with reports of foreign financing terms over extended periods at low interest rates and with little or no down payment.

Even more importantly, the industry viewed U.S. export controls, especially on shipments to former socialist countries, as "one of the major deterrents to the sale of U.S. machine tools overseas."[133] In 1980, the socialist world accounted for 30% of all machine tools consumed in the world and 30% of world imports of machine tools.[134]

Yet because of export regulations, U.S. machine tool builders were unable to exploit this vast market.

The machine tool industry was particularly hard hit by Title IV of the 1974 Trade Act, better known as the Jackson-Vanik amendment tying U.S. export financing to Soviet emigration policy.[135] The combined effect of Jackson-Vanik, the Carter Administration's human rights regulations, and the post-Afghanistan and post-Poland export controls were devastating for machine tool exports. In a 1978 NMTBA survey, 27 of 39 machine tool firms which did business with socialist countries experienced delays and denials in their applications for export licenses. The $58 million of lost sales represented an equivalent of 92% of U.S. exports to socialist countries in 1978.[136] After the first and highly successful U.S. machine tool show in Moscow in 1974, the second show in 1978 turned out to be a complete washout. Orders for U.S. tools were depressed by the lack of Export-Import credit for Soviet purchases, and because four major builders were denied licenses to exhibit NC tools.[137]

Figure 4.10 demonstrates that while socialist machine tool imports skyrocketed, the supplies from U.S. builders dwindled. While socialist countries represented 30% of world machine tool consumption and world machine tool imports, the U.S. supplied only 1.4% of their imports and 0.5% of their total consumption needs in 1980.[138] Between 1975 and 1980, U.S. shipments to the Soviet Union deteriorated 71% and the Soviet Union dropped from the number one U.S. machine tool export market to number six.[139] Figure 4.11 and Table 4.8 show that while the U.S. position in the socialist market collapsed, the other major machine tool suppliers maintained or bolstered their positions.

Figure 4.10

U.S. Share of Total Machine Tool Imports by Socialist Markets

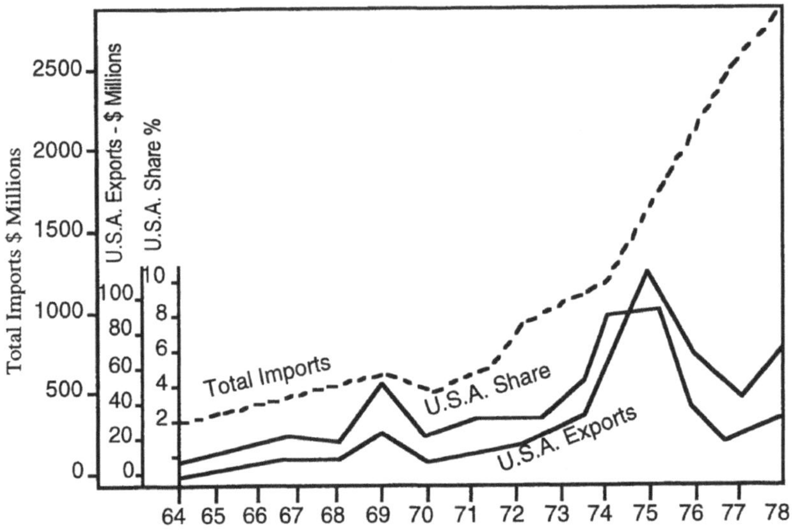

Sources: *American Machinist*, U.S., Bureau of the Census

Figure 4.11

Eastern Bloc Machine Tool Imports from
Six Leading Non-Communist Nations

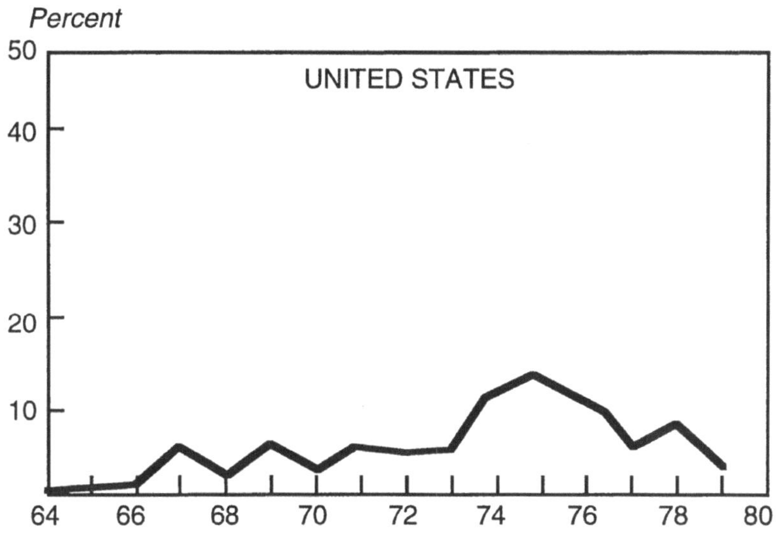

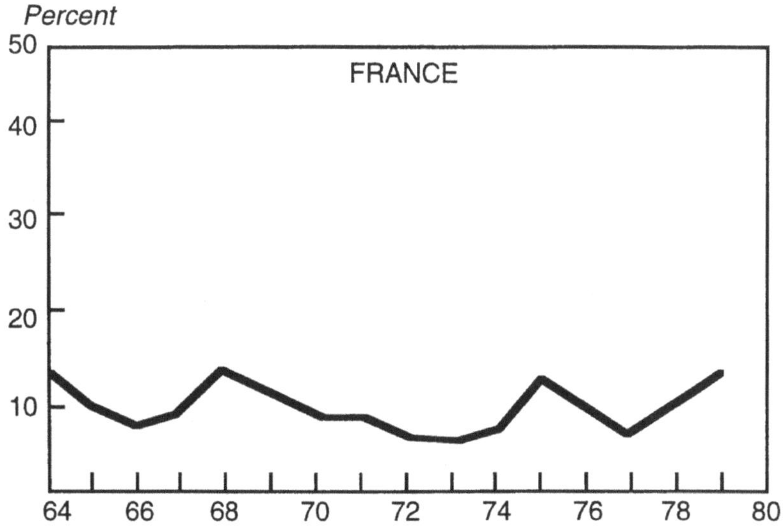

(continued)

Figure 4.11 (continued)

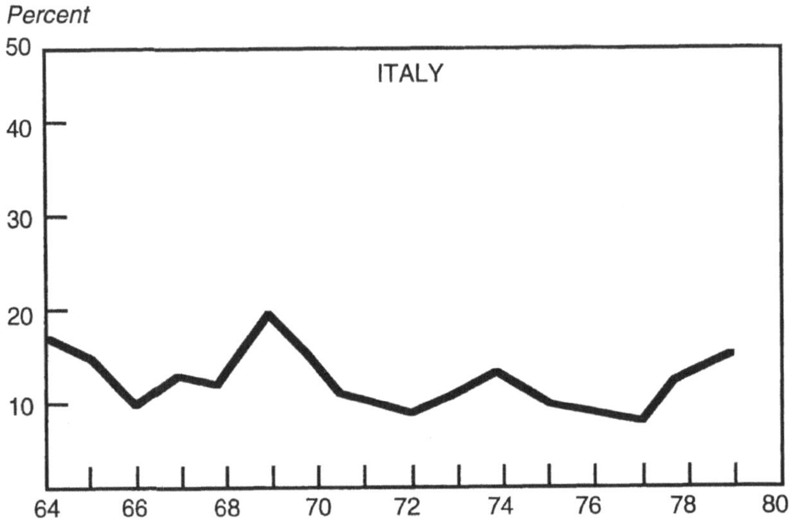

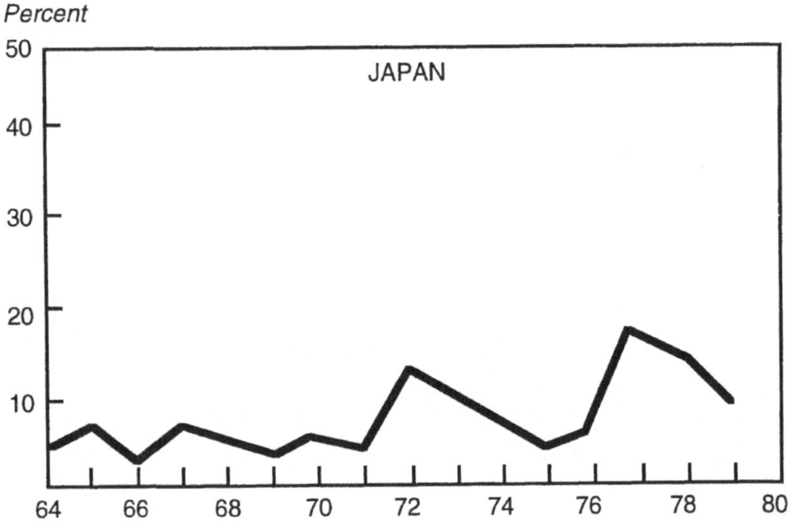

(continued)

Figure 4.11 (continued)

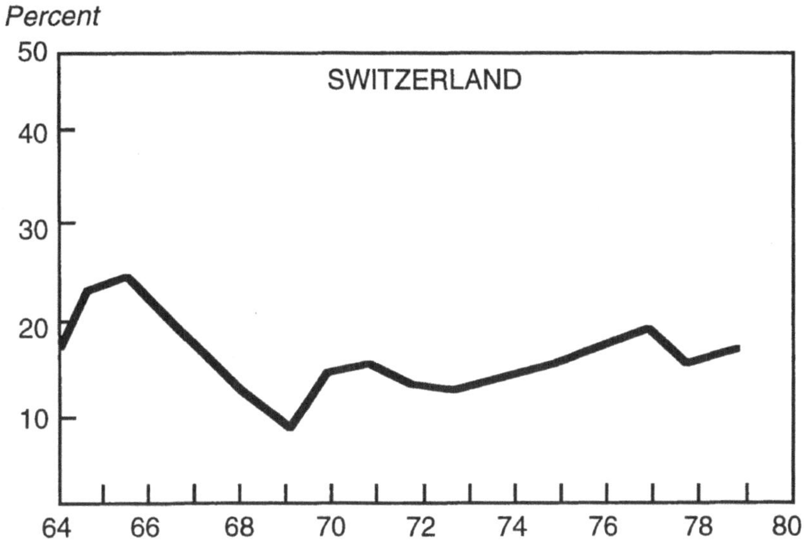

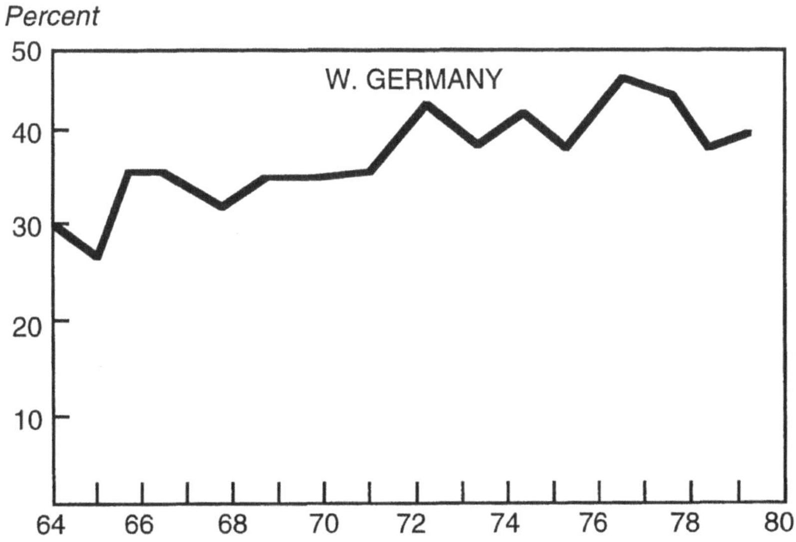

Source: see Table 4.8

Table 4.8

Eastern Bloc(a) Machine Tool Imports from 14 Major Non-Communist Trading Nations(b)
Total and Selected Nations, 1962 to Date (Values in Millions of U.S. Dollars)

		United States		France		West Germany (e)		Italy		Japan		Switzerland	
Year	Total ($)	Value ($)	% (c)	Value ($)	% (c)	Value ($)	% (c)	Value ($)	% (c)	Value ($)	% (c)	Value ($)	% (c)
1962	47.4	.5	1.1	4.0	8.4	16.2	34.2	11.3	23.8	1.0	2.1	5.7	12.0
1963	67.4	.3	.4	6.3	9.3	19.0	28.2	12.9	19.1	2.8	4.2	7.7	11.4
1964	68.7	.2	.3	9.0	13.1	19.4	28.2	12.0	17.5	3.1	4.5	10.9	15.9
1965	60.9	.1	.2	6.2	10.2	16.0	26.3	8.5	14.0	4.0	6.6	14.1	23.2
1966	71.3	.5	.7	5.5	7.7	24.6	34.5	6.3	8.8	1.7	2.4	17.7	24.8
1967	93.6	4.3	4.6	10.5	11.2	32.8	35.0	11.6	12.4	6.1	6.5	16.5	17.6
1968	162.7	5.7	3.5	27.9	17.1	49.6	30.5	19.1	11.7	7.6	4.7	17.8	10.9
1969	307.3	18.1	5.9	40.9	13.3	109.3	35.6	61.1	19.9	9.3	3.0	28.7	9.3

(continued)

Table 4.8 (continued)

		United States		France		West Germany (e)		Italy		Japan		Switzerland	
Year	Total ($)	Value ($)	% (c)	Value ($)	% (c)	Value ($)	% (c)	Value ($)	% (c)	Value ($)	% (c)	Value ($)	% (c)
1970	257.9	8.0	3.1	24.2	9.4	94.2	36.5	40.3	15.6	12.8	5.0	36.5	14.2
1971	201.2	14.8	7.4	19.8	9.8	75.5	37.5	18.0	8.9	10.2	5.1	29.3	14.6
1972	385.5	21.5	5.6	25.4	6.6	164.3	42.6	28.3	7.3	51.2	13.3	43.6	11.3
1973	597.8	32.6	5.5	36.5	6.1	235.7	39.4	48.6	8.1	63.9	10.7	71.5	12.0
1974	727.4	72.2	9.9	48.6	6.7	309.2	42.5	74.2	10.2	50.4	6.9	89.9	12.4
1975	899.6	102.1	11.3	82.1	10.2	355.2	39.5	77.6	8.6	38.7	4.3	125.9	14.0
1976	976.8	63.8	9.8	76.6	7.8	447.9	45.9	79.6	8.1	48.3	4.9	138.0	14.1
1977	1127.8	46.6(d)	4.1	45.6	4.0	504.6	44.7	83.7	7.4	153.7	13.6	144.9	12.8
1978	1331.3	63.3(e)	4.8	116.7	8.8	536.8	40.3	150.0	11.3	165.9(f)	12.5	145.9	11.0
1979	1243.8	18.5(d)	1.5	141.9	11.4	534.9	43.0	155.6	12.5	111.8	9.8	145.9	11.7

(continued)

Table 4.8 (continued)

NA - Not Available

(a) Bulgaria, Czechoslovakia, E. Germany, Hungary, Poland, Romania, and U.S.S.R.
(b) United States, Austria, Belgium-Luxembourg, Canada, Denmark, France, Federal Republic of Germany, Italy, Japan. Netherlands, Norway, Sweden, Switzerland, and United Kingdom.
(c) Percent of total Eastern Bloc machine tool imports from non-communist countries.
(d) Does not include Bulgarian imports.
(e) Does not include East German imports.
(f) Does not include Czechoslovakian imports.

Source: U.S. Dept. of Commerce, "Market Share Reports," Commodity Series, Machine Tools for Working Metals, and NMTBA

The erosion of the industry's export performance became all the more ominous in light of the state of the domestic machine tool business in the 1980s. A good portion of the decline in the industry's export dependence ratio can be attributable to the robustness of domestic machine tool demand. Fueled by automotive tool needs, between 1975 and 1979 domestic machine tool orders and order backlogs rose 382% and 303%, respectively (in current dollars).[140] As discussed earlier in the chapter, when domestic business booms, the relative importance of exports decline. This occurred in the early seventies as well as the late seventies. Hence, much of the decline in export dependence since 1975 was to be expected.

Yet when orders and backlogs plummeted in 1980 and through 1982, one would have expected the industry's reliance on exports to increase. For when the domestic market weakens, export dependence tends to heighten as was the case in the early sixties, in 1971, and in 1975. But such was not the case in the recession of the early 1980s. Instead of recovering, export dependence ratios continued to decline along with the domestic market. Instead of acting as a natural buffer against protectionism during bad economic times, the constraints of export dependence eroded at the same time that motivations for protection — import penetration, hegemonic decline, economic recession — came to a head. The last time such a vicious combination occurred was in the late sixties. And, of course, that period manifested the industry's most protectionist behavior.

Beyond the weakening of export performance, the structure of U.S. trade in machine tools made it very difficult for export dependence to act as an effective constraint against protectionism. In particular, the one country was creating most of the import problems and against which virtually all protectionist sentiments were directed against, Japan, represented a market the U.S. industry was *not* dependent upon for machine tool sales.

Figure 4.12 demonstrates that through the seventies, U.S. exports to Japan stagnated as much as imports from Japan skyrocketed. The bilateral trade balance went into deficit for the U.S. in 1975 and the ratio of U.S. imports to U.S. exports came to stand at 13 to 1 ($687.5 million versus $54.2 million in 1981). The bilateral trade balance becomes all the more lopsided when considering only the all-important NC machine tools (see Table 4.9). The overall machine tool trade balance of 13 to 1 in favor of Japan grows to 87 to 1 for all NC tools and explodes yet another 100 times worse (871 to 1) in NC lathes! The balance literally collapses on the Japanese side in machining center trade where in face of nearly 2000 Japanese sales in the U.S. market,

U.S. builders failed to make a single machining center sale in the Japanese market in 1981.

Figure 4.12

U.S. Trade with Japan in Machine Tools

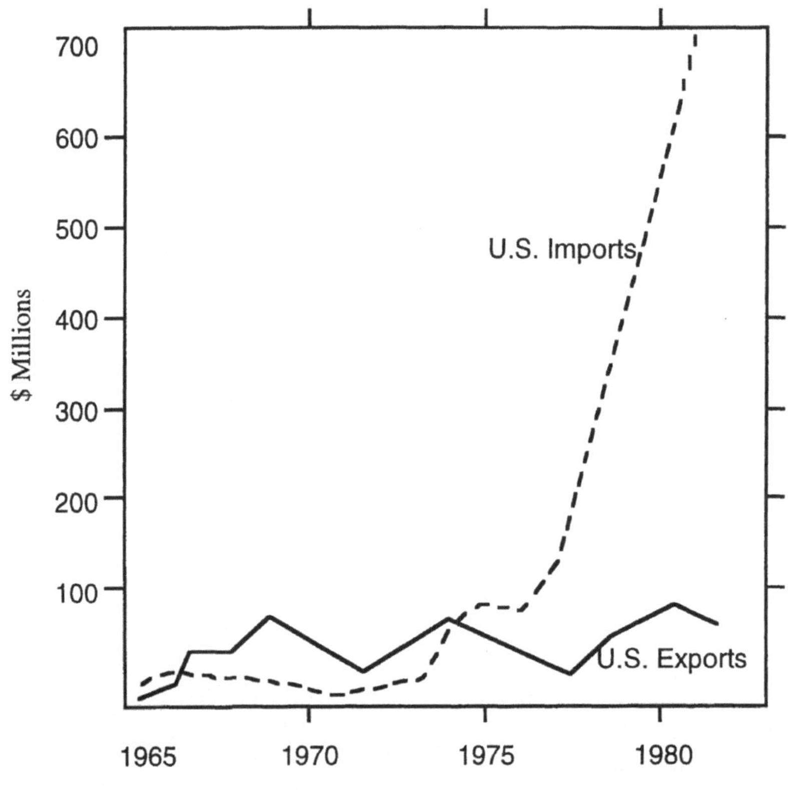

Sources: U.S., Bureau of the Census, *U.S. Exports: Schedule B; U.S. Imports for Consumption: TSUSA Schedule*

Table 4.9

U.S. Trade with Japan in Numerically Controlled Machine Tools, 1981

	U.S. Exports to Japan		U.S. Imports from Japan	
	units	value ($000)	units	value ($000)
Metal-Cutting	27	6706	5451	473,379
Boring	2	553	39	7095
Drilling	1	81	19	518
Grinding	20	5764	98	6648
Lathes	3	297	3211	258,592
Milling	0	0	131	16,303
Machining	0	0	1944	188,851
NEC	1	11	9	372
Metal-Forming	37	2658	176	17,698
Punching	16	1227	162	16,305
Bending	21	1429	5	380
NEC	0	0	9	1013
Total	64	9364	5627	491,077

Sources: NMTBA; U.S., Bureau of the Census, *U.S. Exports: Schedule B*, EM 522; *U.S. Imports for Consumption: TSUSA Schedule*, IM 146

The market consequences of these trends are presented in Figure 4.13 and Table 4.10. The proportion of U.S. exports destined for Japan declined from an approximately 20% share in the late sixties-early seventies to a meager 5% share in 1981. Far from growing dependent upon sales to the Japanese market, the U.S. machine tool industry effectively disengaged from it. At one point in 1971, a respectable 5% of all U.S. machine tool sales were made in Japan. Ten years later hardly 1% of U.S. business was conducted with the Japanese.

At the same time, of course, Japanese builders were taking increasing shares of the U.S. machine tool market. With shipments to the U.S. equaling those of the rest of the top ten foreign machine tool suppliers to the U.S. combined,[141] Japan in 1981 came to capture over 12% of the U.S. market on its own.[142]

The situation, then, had become increasingly intolerable for the U.S. machine tool industry. The 1980s brought economic recession on top of ever-intensifying foreign competition. And what had proven to be an effective deterrent against protectionism, export dependence, weakened at a time one would expect resurgence. On top of all this, the market of the one foreign supplier that was causing most of the import problem meant literally nothing to the fortunes of the U.S. industry. With virtually no stake in the Japanese market, U.S. builders could opt for protectionist policies with little to fear or lose from Japanese retaliation. The negligible position of the U.S. in the Japanese market could not get much worse.

We had the makings here of a tragic combination: high motivations for protectionism simultaneous with reduced constraints against such behavior. The last time such a combination manifested, in 1968, the results were disastrous.

Indeed, protectionist pressures within the machine tool industry mounted in the first six months of 1982 to levels not seen in over a decade. These pressures had four major foci: U.S. defense procurement, patent infringement, the Houdaille trade petition, and national security.

Figure 4.13

U.S.-Japan Machine Tool Trade Ratios

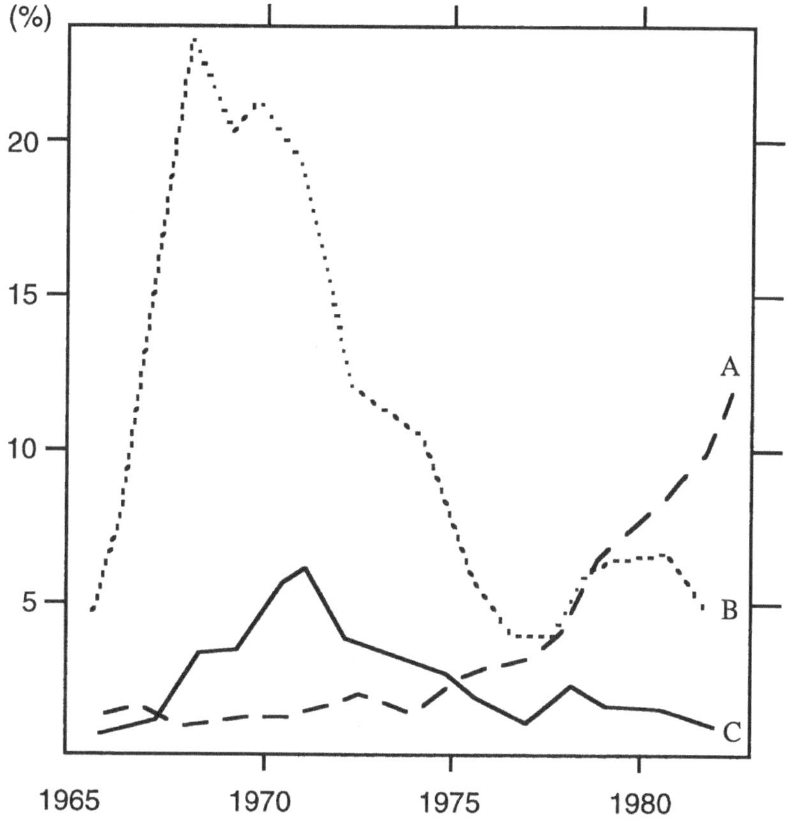

A Japanese Import Penetration of the U.S. Market
B Japanese Share of U.S. Exports
C U.S. export Dependence on the Japanese Market

Sources: U.S., Bureau of the Census, *U.S. Exports: Schedule B,* EM 522; *U.S. Imports for Consumption: TSUSA Schedule,* IM 146; U.S., Bureau of Economic Analysis, *Business Statistics; Survey of Current Business.*

Table 4.10

U.S.-Japan Machine Tool Trade Ratios (%)

	Japanese Import Penetration of the U.S. Market	Japanese Share of U.S. Exports	U.S. Export Dependence on the Japanese Market
1966	1.12	0.66	5.00
1967	1.53	1.17	8.98
1968	1.20	2.95	22.64
1969	1.23	3.13	19.72
1970	1.23	4.34	20.52
1971	1.30	5.02	18.72
1972	1.77	3.07	12.04
1973	1.73	2.68	11.47
1974	2.64	2.42	11.06
1975	2.94	1.57	6.78
1976	3.82	1.31	4.94
1977	4.84	0.97	4.88
1978	7.10	1.78	6.96
1979	8.28	1.25	7.44
1980	9.63	1.26	7.53
1981	12.39	1.06	5.27

Sources: see Figure 4.13

As a first blow against foreign imports, the industry staked out a definitive position against the purchase of foreign machine tools for U.S. defense requirements. The industry has, in fact, called for strong Buy America Act provisions for military purchases since the 1950s.[143] These sentiments intensified with the celebrated 1978 Defense Department purchase of Mitsubishi machining centers for manufacturing the XM-l tank.[144] The industry's position on this issue hardened further in May 1982 — a portentous month as will be established — when the NMTBA launched a major public and political campaign to deny foreign machine tools from consideration for military needs. This campaign was initiated by a full page ad on May 11 in the Washington Post.

The industry also took steps to crack down on those foreign builders who duplicated the designs of standard U.S. machine tools. By offering such "copies" at prices 50% below those of U.S. originals, such foreign builders, largely in Taiwan and South Korea, undercut U.S. sales both at home and abroad.[145] Since the mid-1970s a handful of U.S. machine tool builders have sued foreign firms for various patent infringements.[146] Yet in contrast to these isolated actions, the NMTBA initiated its own industry-wide effort against foreign copiers in May 1982, the same month the association launched its campaign against the use of foreign machine tools for military needs. In actions similar to NMTBA moves against alleged dumping in late 1977, the association has solicited from its members any information on foreign copies so that it might be in a position to seek import relief.[147] In principle and often in fact, such actions constitute legitimate counter-measures against unfair trade practices and do not represent protectionist sentiments. Yet in being initiated during a period of other protectionist pressures, the industry's "justified" complaints against foreign copies may very well have been a guise for more malign purposes.

The two trade actions discussed above were fairly limited in scope. The first is relevant for only one segment of the machine tool market: defense procurement. The second is aimed only at developing machine tool builders in South Korea and Taiwan and not against the formidable U.S. rivals in Western Europe and Japan. Indeed, persons inside and outside the industry could and did argue that protectionism did not characterize the industry's positions on military procurement and foreign copies. Such an argument *cannot* be made for the industry's next two trade actions — actions that explicitly sought the protection of the lucrative civilian as well as military market from all foreign and especially Japanese products.

In a third trade action of May 1982, a major machine tool firm, Houdaille Industries, Inc., filed a petition with the USTR charging that collusion among Japanese builders along with the Japanese government on machine tool market shares and prices provided Japan unfair advantages in penetrating the U.S. market. For relief, Houdaille requested that the President suspend investment tax credits for purchases of Japanese machine tools — a move that was projected to increase the prices of Japanese machines by 15%.[148]

The final trade action to be reported was the most serious and pervasive of the four discussed. Early in 1982, the NMTBA's Governmental Relations Committee directed its staff to study options for dealing with the import problem. In April the committee concluded that import relief could best be sought under a seldom exercised provision, Section 232, of the Trade Expansion Act of 1962. Section 232 requires the Commerce Department, upon demand, to investigate threats to national security from imports. The industry contended that the import for foreign machine tools "places the defense and security of the nation in grave jeopardy, because it tends to create an enduring dependence of our defense industry on foreign sources [of machine tools] which are essential for the manufacture of defense systems."[149]

It is important to note that this argument has increasingly been accepted in Congress, Executive offices with trade policy responsibilities, the Defense Department, and the CIA.[150] At the time of this writing, the NMTBA had yet to file a 232 petition, but it was deeply engaged in studying such a move. Such a motion could easily result in the imposition of the severest of trade barriers to protect the machine tool industry.

Once again we see the effect of changing levels of export dependence on machine tool trade policy. The heightening of export dependence in the early 1970s led to a decade of liberal trade policy. The weakening of the constraints of export dependence in the 1950s and 1960s permitted the ascendance of protectionism that culminated with the 1968 proposal for import surcharges. With the onset of the 1980s, the industry's advocacy of liberal trade was once again abated as export dependence levels diminished in face of mounting international and domestic ills.

CONCLUSION

The machine tool industry serves as a preeminent "hypothesis-testing" case study for the export dependence thesis. For a thesis that seeks to explain liberal trade policy, the machine tool case is significant for all the reasons *not* to have expected the industry to adopt such policy views in the 1970s. Chapter III demonstrated the severe import penetration and hegemonic decline that the industry has experienced. Conventional analysts would be quick to point to such trends in presuming a protectionist-ridden industry. Other reasons for not expecting the industry to have adopted liberal trade views include the industry's minimal foreign direct investment, its cyclical economic performance, and its lack of technological innovation and loss of technological leadership over foreign builders. On top of all this, the industry is considered vital to national security.

Indeed, the machine tool case was chosen precisely because all the conventional indicators point heavily towards a protectionist outcome. Only the industry's high levels of export dependence in the 1970s stood as a bulwark against such an outcome. If, despite all the countervailing pressures, the industry's dependence upon exports acted to effectively constrain machine tool protectionism, then the findings take on immense significance. If the constraints of export dependence are operative even under the worst of conditions, then its impact must be considered substantial. It might have been easier to make the case for export dependence with an industry where conditions were less difficult and the countervailing pressures less evident. The findings, however, would have been less meaningful.

The machine tool industry also proves to be a "hypothesis-confirming" case study. We have directly linked levels of export dependence and the industry's trade policy views over four decades. When the constraints of export dependence were low in the 1950s and 1960s, machine tool protectionism flourished. When the constraints heightened in the 1970s, the industry adopted liberal commercial policy positions — again, despite the mounting countervailing pressures. With the 1980s we witnessed the concomitant deterioration of machine tool export dependence and liberal trade policy.

The prominence of this explanation is enhanced given the inadequacy of alternative explanations. The industry pursued a liberal trade policy in the 1970s without the benefit of significant foreign direct investment. At the same time, the industry's hegemonic and relative technological positions significantly deteriorated while foreign import penetration rose dramatically. An economic explanation based upon the domestic business cycle can neither explain the machine tool protectionism sustained during the growth period between 1958 and 1966, the adoption of liberal policy during the bust of 1971, nor the maintenance of liberal policy through the 1974-75 recession. Only the export dependence thesis can consistently account for machine tool commercial policy from the 1950s into the 1980s.

This is not to argue that pressures emanating from import penetration, hegemonic decline, economic recession, or deteriorating technological advantage have absolutely no impact on machine tool trade policy. Indeed, they always have an impact. But an important conclusion of this study is that such motivations for protectionism can be offset by the constraints of export dependence. Such motive forces have flourished only when they have been unimpeded by export dependence. Trade policy cannot be explained solely on the basis of motivations for protectionism.

Our analysis of the machine tool industry offers three mini-case studies confirming the export dependence thesis: the 1950s-1960s, the 1970s, and the early 1980s. In each case varying levels of export dependence made a differential impact on machine tool trade policy, and at no time does the industry's behavior deviate from hypothesized expectations. Taken together, the three episodes not only link export dependence to trade policy but establish the dynamism of that relationship.

NOTES

1 *American Metal Market/Metalworking News*, February 13, 1978, p. 8; February 26, 1979, p. 8; July 28, 1980, p. 8; *Financial World*, January 1, 1979, p. 31; *Iron Age*, August 18, 1980, p. 67.

2 For similar conclusions see Bernard Real, *Technical Change and Economic Policy: Science and Technology in the New Economic and Social Context, Sector Report, The Machine Tool Industry* (Paris: OECD, August 1, 1980).

3 Otto Hintz, *et. al., Machine Tool Industry Final Report*, U.S. Army Industrial Base Engineering Activity, Rock Island, Illinois, November 1978, p. 36; National Science Foundation, "National Patterns of Science and Technology Resources, 1980," NSF 80-308, p. 46; both cited in Judith Reppy, "The Role of the Air Force in the Development of Numerically Controlled Machine Tools," 1981 (mimeo); *Economist*, March 6, 1982, pp. 95-6.

4 Wilbur F. Monroe, *Japanese Exports to the United States: Analysis of "Import Pull" and "Export Push" Factors* (Washington: U.S.-Japan Trade Council, June 2, 1978), p. 159.

5 *Industry Week*, September 1, 1980.

6 *New York Times*, February 26, 1978.

7 *Industry Week*, September 1, 1980.

8 Real, *Technical Change*, p. 28.

9 U.S. Bureau of the Census, "Current Industrial Reports, Series MQ-35W, Metalworking Machinery," Japan, Ministry of International Trade and Industry.

10 Real, *Technical Change*, p. 62.

11 See footnote no. 9.

12 Real, *Technical Change*, p. 62.

13 *Ibid.*, , pp. 25, 56.

[14] Since the early 1970s, in one of the most important cross-national marketing arrangements in numerical controls, Fanuc controls have been marketed throughout Western Europe by Siemens. Direct sales of NC units in Europe through Siemens and indirect sales to the U.S. (i.e., Japanese machine tools equipped with Fanuc controls) serve as the basis for Fanuc's domination of the world market. In 1975, the two firms entered into a U.S. joint venture, under the name of General Numeric, to produce controls in the U.S.. In 1981, General Numeric was the fifth largest NC producer in the U.S. (*American Metal Market/Metalworking News*, May 18, 1981, pp. 24-29).

[15] *American Machinist*, p. 37.

[16] *Iron Age*, December 1, 1980.

[17] *Ibid.*, December 17, 1979.

[18] *American Metal Market/Metalworking News*, September 28, 1981, p. 10.

[19] *Iron Age*, September 18, 1978, p. 11; December 17, 1979.

[20] *Ibid.*, September 18, 1978, p. 11; May 19, 1980; American Machinist, May 1980, pp. 167-182.

[21] *American Machinist*, June 1981.

[22] *Ibid.*, February 1981, p. 98.

[23] *Iron Age*, November 20, 1978, p. 91.

[24] *Ibid.*, November 24, 1980, p. 119.

[25] *American Metalworking Market/Metalworking News*, July 13, 198, pp. 8.

[26] Real, *Technical Change*, p. 21.

[27] *Iron Age*, August 27, 1979, p. 21.

[28] *American Metal Market/Metalworking News*, January 5, 1981.

29 *Steel*, February 3, 1958, pp. 66-7; April 21, 1958, pp. 57-8; August 11, 1958, p. 77.

30 U.S., Bureau of the Census, *Census of Manufactures*; *Annual Survey of Manufactures*.

31 *Iron Age*, March 11, 1971, p. 35; *Barron's*, December 14, 1970, p. 15.

32 *Industry Week*, February 10 1975, p. 61; *American Machinist*, January 1976.

33 *Iron Age*, March 6, 1978, p. 9; October 8, 1979, p. 13; January 14, 1980; *American Machinist*, November 1979, p. 91.

34 *Standard & Poor's Industry Survey*, October 1981, p. M25.

35 *Ibid.*, July 1981; October 1981.

36 U.S., Bureau of Economic Analysis, *Survey of Current Business*.

37 *Ibid.*

38 *Standard & Poor's Industry Survey*, October 1981; *New York Times*, May 31, 1981, p. Dl,4.

39 *American Metal Market/Metalworking News*, June 29, 1981, p. 1; January 25, 1982, p. 1; *Iron Age*, December 16, 1981, p. 17.

40 *American Metal Market/Metalworking News*, October 5, 1981, pp. 12, 19.

41 *Ibid.*

42 *New York Times*, April 26, 1982, p. Dl.

43 *Ibid.*

44 The ramifications of the early 1980s recession are discussed in a later section.

45 This and the following paragraph are based on U.S., Congress, House, Committee on Ways and Means, *Trade Agreements Extension*, hearings, part 2,

January 17, 1955, 84th Congress, First Session; and U.S. Congress, Senate, Select Committee on Small Business, *Machine Tool Programs*, hearings, February 7, 1956, 84th Congress, Second Session.

[46] U.S., Bureau of Economic Analysis, *Business Statistics*.

[47] U.S., Congress, House, Committee on Ways and Means, *Trade Expansion Act of 1962*, hearings, March 12, 1962, 87th Congress, 2nd Session, p. 1294.

[48] U.S., Congress, House, Select Committee on Export Control, *Investigation and Study of the Administration, Operation, and Enforcement of the Export Control Act of 1949, and Related Acts*, hearings, October 25, 1961, 87th Congress, 1st Session, p. 487.

[49] U.S., Congress, *Trade Expansion Act of 1962*, March 12, 1962, p. 1213.

[50] *Ibid.*, p. 1294.

[51] NMTBA, "Industry Estimates of Machine Tool New Orders, Cancellations, Shipments and Backlog."

[52] These phrases are taken from Kenneth N. Waltz, *Man, the State, and War: A Theoretical Analysis* (New York: Columbia University Press, 1959), pp. 232-36.

[53] U.S., Bureau of the Census, "Current Industrial Reports."

[54] U.S., Bureau of the Census, *U.S. Exports: Schedule B Section by Division by Group by Number by Country of Destination by Customs District of Exportation and Method of Transportation*, EM 522.

[55] The following two paragraphs are drawn from U.S., Congress, Senate, Committee on Banking, Housing and Urban Affairs, Subcommittee on International Finance, *Authority for Regulation of Exports - 1972*, hearings, May 13, 1972, 92nd Congress, Second Session, p. 4; U S., Congress, Senate, Committee on Finance, *The Trade Reform Act of 1973*, hearings, part 4, May 26, 1974, p. 1504; and U.S., Congress, House, Committee on International Relations, Subcommittee on International Economic Policy and Trade, *Export Licensing: Cocom List Review Proposals of the United States*, hearings, June 14, 1978, 95th Congress, Second Session, p. 23 unless otherwise indicated.

[56] U.S., Bureau of the Census, *U.S. Exports: Schedule B*; U.S., Congress, Senate, Committee on Banking, Housing and Urban Affairs, Subcommittee on International Finance, *Export Policy*, hearings, part 4, March 20, 1978.

[57] U.S., Bureau of the Census, *U.S. Exports: Schedule B.*

[58] U.S., Congress, *Export Policy*, March 20, 1978, p. 339.

[59] U.S., Congress, Senate, Committee on Banking, Housing and Urban Affairs, Subcommittee on International Trade, *Export Trading Company Act of 1980*, March 17, 1980, 96th Congress, Second Session, p. 8; *American Machinist*, February 1976.

[60] *Annual Report* - Cincinnati Milacron Inc.

[61] See U.S., Congress, *Authority for Regulation of Exports - 1972*, March 13, 1972, pp. 187, 329; U.S., Congress, House, Committee on Ways and Means, Subcommittee on Trade, *Causes and Consequences of the U.S. Trade Deficit and Developing Problems in U.S. Exports*, hearings, November 3, 1977, 95th Congress, First Session; U.S., Congress, House, Committee on Banking, Finance and Urban Affairs, Committee on International Trade, Investment and Monetary Policy, *Export-Import Bank and Trade of China*, hearings, January 26, 1978, 95th Congress, Second Session, p. 76; U.S., Congress, *Export Licensing*, June 14, 1978, p. 17; James A. Gray, "Machine Tool Exports: A Smaller Slice of a Bigger Pie," November 1979, p. 12; U.S., Congress, *Export Trading Company Act of 1980*, March 17, 1980, p. 47.

[62] U.S., Congress, *Trade Agreements Extension*, January 17, 1955, pp. 1936, 1938-39; U.S., Congress, House, Committee on Ways and Means, *Renewal of Trade Agreements Act*, hearings, part 2, March 10, 1958, 85th Congress, Second Session, p. 1747; U.S., Congress, *Investigation and Study*, October 25, 1961, pp. 478, 487; U.S., Congress, House, *Trade Expansion Act of 1962*, March 12, 1962, p. 1286.

[63] *Steel*, July 31, 1972, p. 18.

[64] U.S.. Congress, House, Committee on Banking and Currency, Subcommittee on International Trade, *International Economic Policy*, hearings, April 22, 1974, 93rd Congress, Second Session, p. 649.

[65] *Ibid.*; National Machine Tool Builders Association, *1981-1982 Economic handbook of the Machine Tool Industry* (NMTBA, 1981), p. 77.

[66] U.S., Congress, House, Committee on Ways and Means, Subcommittee on Trade, *Trade with Japan*, hearings, August 26, 1980, 96th Congress, Second Session, p. 165; see also *Iron Age*, November 15, 1976, pp. 9-10.

[67] U.S., Congress, Senate, Committee on Banking, Housing and Urban Affairs, Subcommittee on International Finance, *U.S. Export Control Policy and Extension of the Export Administration Act*, hearings, part 2, March 12, 1979, 96th Congress, First Session, p. 131. See also U.S., Congress, *Causes and Consequences of the U.S. Trade Deficit*, November 3, 1977, pp. 397-98; U.S., Congress, *Export Policy*, March 20, 1978, p. 278; U.S., Congress, House, Committee on Foreign Affairs, Subcommittee on International Economic Policy and Trade, *Extension and Revision of the Export Administration Act of 1969*, hearings and markup, February 15, 1979, 96th Congress, First Session, p. 208.

[68] U.S., Congress, Senate, Committee on Banking, Housing and Urban Affairs, Subcommittee on International Finance, *Extension of the Export Administration Act*, hearings, March 2, 1976, 94th Congress, Second Session, p. 329. See also U.S., Congress, *Causes and Consequences of the U.S. Trade Deficit*, November 3, 1977, p. 396; U.S., Congress, *Export Policy*, March 20, 1978, p. 277; Gray, "Machine Tool Exports," pp. 1-12, emphasis added.

[69] Cited in U.S., Congress, *Machine Tool Programs*, February 7, 1956, p. 164, emphasis added.

[70] *Business Week*, June 20, 1959, p. 187.

[71] U.S., Congress, House, Committee on Ways and Means, *Tariff and Trade Proposals*, hearings, part 9, June 4, 1970, 91st Congress, Second Session, p. 2549.

[72] Statement by Philip D. Grier, Cincinnati Milacron President, cited in *Steel*, July 1, 1968, p. 34.

[73] U.S., Congress, *Tariff and Trade Proposals*, June 4, 1970, p. 2549.

[74] *Ibid.*

[75] *Ibid.*

[76] *Ibid.*, p. 2551.

[77] U.S., Congress, House, Committee on Ways and Means, *Trade Reform*, hearings, part 3, May 14, 1975, 93rd Congress, First Session, p. 803, emphasis added. See a similar statement by the President of Cross Company and NMTBA First Vice President in U.S., Congress, *The Trade Reform Act of 1973*, March 26, 1974, p. 1505.

[78] U.S., Congress, *Trade with Japan* , August 26, 1980, p. 167, emphasis added.

[79] U.S. Congress, *Machine Tool Programs*, February 7, 1956, p. 170; U.S., Congress, House, Committee on Education and Labor, Subcommittee on the Impact of Imports and Exports on American Employment, *Impact of Imports and Exports on Employment*, hearings, part 8, November 27, 1961, 87th Congress, First and Second Sessions, p. 808; U.S., Congress, House, Committee on Ways and Means, *Foreign Trade and Tariff Proposals*, hearings, part 7, June 21, 1968, 90th Congress, Second Session, p. 2870; U.S., Congress, *Tariff and Trade Proposals*, June 4, 1970, p. 2543; U.S., Congress, *The Trade Reform Act of 1973*, March 26, 1974, pp. 1505-6; "Report of the Industry Sector Advisory committee on Machine Tools, Other Metalworking Equipment and Other Nonelectrical Machinery (ISAC 17) to the President, the Congress, and the Special Representative for Trade Negotiations, April 18, 1979," in U.S., Senate, Committee on Finance, Subcommittee on International Trade, *Private Advisory Committee Reports on the Tokyo Round of Multilateral Trade Negotiations*, August 1979.

[80] *American Metal Market/Metalworking News*, February 14, 1972, pp. 1, 23.

[81] *Ibid.*, December 12, 1977, p. 2.

[82] *Ibid.*, January 23, 1978, p. 14.

[83] *Ibid.*, June 18, 1979, p. 4; June 25, 1979, p. 6.

[84] *Ibid.*, February 22, 1982, p. 4a.

[85] *Ibid.*, May 31, 1982, p. 1.

[86] U.S., Congress, *Foreign Trade and Tariff Proposals*, June 21, 1968, p. 2870; U.S., Congress, Senate, Committee on Banking, Housing and Urban Affairs, Subcommittee on International Finance, *Export-Import Bank of the United States*, hearings, October 29, 1973, 93rd Congress, First Session, p. 221.

87 U.S., Congress, *Causes and Consequences of the U.S. Trade Deficit*, November 3, 1977, pp. 415-19; U.S., Congress, Senate, Committee on Banking, Housing, and Urban Affairs, Subcommittee on International Finance, *Amending the Export-Import Bank Act and the Trade Act*, hearings, July 12, 1979, 96th Congress, First Session; U.S., Congress, House, Committee on Ways and Means, Subcommittee on Trade, *United States-China Trade Agreement*, hearings, November 1, 1979, 96th Congress, First Session, p. 341, U.S., Congress, Senate, Committee on Finance, Subcommittee on International Trade, *Agreement on Trade Relations Between the United States and the People's Republic of China*, hearings, November 15, 1979, 96th Congress, First Session, p. 72.

88 U.S., Congress, *Investigation and Study*, October 25, 1961; *Foreign Trade and Tariff Proposals*, June 21, 1968, p. 2870; *Tariff and Trade Proposals*, June 4, 1970, pp. 2543-44; *Authority for Regulation of Exports*, March 13, 1972, p. 184; *International Economic Policy*, April 22, 1974, p. 529; *Extension of the Export Administration Act*, March 22, 1976, p. 329; U.S., Congress, House, Committee on International Relations, *Extension of the Export Administration Act of 1969*, hearings, part 1, June 8, 1976, 94th Congress, Second Session; U.S., Congress, *Causes and Consequences of the U.S. Trade Deficit*, November 3, 1977, pp. 409-15; *Export Licensing*, June 14, 1978, pp. 22-43; *Extension and Revision of the Export Administration Act of 1969*, February 15, 1979, p. 221; *U.S. Export Control Policy*, March 12, 1979; "Statement by National Machine Tool Builders' Association before the Subcommittee on Trade, Committee on Ways and Means, U.S. House of Representatives, December 15, 1981," pp. 18-20 (mimeo).

89 U.S., Congress, *The Trade Reform Act of 1973*, March 26, 1974, p. 1507; *International Economic Policy*, April 24, 1974; *Causes and Consequences of the U.S. Trade Deficit*, November 3, 1977, p. 403; *Export-Import Bank and Trade with China*, January 26, 1978; *Export Policy*, March 20, 1978; *Amending the Export-Import Bank Act*, July 12, 1979; U.S., Congress, Senate, Select Committee on Small Business, *Small Business Participation in U.S. Exports*, hearings, September 25, 1979, 96th Congress, First Session.

90 U.S., Congress, *Machine Tool Programs*, February 7, 1956, U.S., Congress, House, Committee on Banking and Currency, *Export-Import Bank Act Extension*, hearings, February 25, 1963, 88th Congress, First Session; U.S., Congress, *Foreign Trade and Tariff Proposals*, June 21, 1968, p. 2870; Tariff and Trade Proposals, June 4, 1970, pp. 2544-45; U.S., Congress, Senate, Committee on Banking, Housing and Urban Affairs, Subcommittee on International Finance, *Amend the Export-Import Bank Act of 1945*, hearings, March 8, 1971, 92nd Congress, First Session, pp. 44-45; U.S., Congress,

Export-Import Bank of the United States, October 29, 1973, p. 218; *International Economic Policy*, April 22, 1974; *Causes and Consequences of the U.S. Trade Deficit*, November 3, 1977, pp. 403-6; *Export-Import Bank of Trade with China*, January 26, 1978; *Export Policy*, March 20, 1978, pp. 276-82; *Amending the Export-Import Bank Act*, July 12, 1979, p. 165; *Export Trading Company Act of 1980*, March 17, 1980; "Statement by the National Machine Tool Builders' Association," December 15, 1981, pp. 16-18.

91 U.S., Congress, *Foreign Trade and Tariff Proposals*, June 21, 1968, pp. 2870-71; U.S., Congress, Senate, Committee on Commerce, Subcommittee on Foreign Commerce and Tourism, *Proposed Reductions in the Export Promotion Budget and Changes in Fees for Commerce Department Services*, hearings, April 28, 1975, 94th Congress, First Session; U.S., Congress, *Causes and Consequences of the U.S. Trade Deficit*, November 3, 1977, pp. 419-21; *Small Business Participation in U.S. Exports*, September 25, 1979; *Export Trading Company Act of 1980*, March 17, 1980.

92 U.S., Congress, *Foreign Trade and Tariff Proposals*, June 21, 1968, p. 2869; *Tariff and Trade Proposals*, June 4, 1970, pp. 2542-43, 2545; *Causes and Consequences of the U.S. Trade Deficit*, November 3, 1977, pp. 421-23; U.S., Congress, Senate Committee on Finance, *Revenue Act of 1978*, hearings, part 3, August 17, 1978, 95th Congress, Second Session; "Statement by the National Machine Tool Builders Association," December 15, 1981, pp. 16-18.

93 U.S., Congress, *Small Business Participation in U.S. Exports*, September 25, 1979; *Export Trading Company Act of 1980*, March 17, 1980; U.S., Congress, House, Committee on Banking, Finance and Urban Affairs, Subcommittee on International Trade, *Investment and Monetary Policy, Export Trading Companies Legislation*, hearings, July 1, 1980, 96th Congress, Second Session; U.S., Congress, Senate, Committee on Banking, Housing and Urban Affairs, Subcommittee on International Finance and Monetary Policy, *Export Trading Company Act of 1981*, hearings, February 17, 1981, 97th Congress, First Session; "Statement by Wayne R. Moore, President and Chief Executive Officer, Moore Special Tool Company, Inc. before the Subcommittee on Banking, Housing and Urban Affairs, U.S. House of Representatives, April 22, 1982" (mimeo).

94 On R&D, see U.S., Congress, House, *Trade Expansion Act of 1962*, March 12, 1962; *Trade with Japan*, August 26, 1980, p. 168. On technical training, see *American Machinist*, December 1981, p. 49. On depreciation and investment tax credits, see U.S., Congress, Senate, Select Committee on Small Business, *Machine Tool Shortages*, hearings, February 27, 1952, 82nd Congress, Second Session, p. 529; U.S., Congress, *Machine Tool Programs*, February 7, 1956; *Investigation and Study*, October 25, 1961; *Impact of Imports*

and Exports on Employment, November 27, 1961; *Trade Expansion Act of 1962*, March 12, 1962, p. 2871; *Foreign Trade and Tariff Proposals*, June 21, 1968; *Tariff and Trade Proposals*, June 4, 1970, p. 2540; *The Trade Reform Act of 1973*, March 26, 1974, p. 1508; *Revenue Act of 1978*, August 17, 1978; U.S., Congress, Senate, Committee on Finance, Subcommittee on Taxation and Debt Management Generally, *Tax Incentives for Exports*, hearings, June 18, 1979, 96th Congress, First Session; U.S., Congress, House, Committee on Ways and Means, *Advisability of a Tax Reduction in 1980 Effective for 1981*, hearings, part 1, July 22, 1980, 96th Congress, Second Session; U.S., Congress, *Trade with Japan*, August 26, 1980, pp. 167-68; U.S., Congress, House, Committee on Ways and Means, *Tax Aspects of the President's Economic Program*, hearings, part 2, February 24, 1981, 97th Congress, First Session.

95 U.S., Congress, *Tax Aspects of the President's Economic Program*, February 24, 1981, p. 1316.

96 James A. Gray, "The American Machine Tool Industry," February 12, 1982, p. 4 (mimeo).

97 U.S., Congress, *Causes and Consequences of the U.S. Trade Deficit*, November 3, 1977, p. 402, also in *Export Policy*, March 20, 1978, p. 279.

98 U.S., Congress, *Export Trading Company Act of 1980*, March 17, 1980, p. 10; *Export Trading Company Act of 1981*, February 17, 1981.

99 *American Metal Market/Metalworking News*, May 1978, p. 93; *Iron Age*, July 17, 1978, p. 15.

100 *Iron Age*, March 3, 1980, pp. 34-35.

101 *American Metal Market/Metalworking News*, August 31, 1981, p. 20.

102 *Industry Week*, September 7, 1981, p. 140; *American Metal Market/Metalworking News*, September 14, 1981, p. 14.

103 *American Metal Market/Metalworking News*, December 14, 1981, p. 17; January 1, 1982, p. 17.

104 *Iron Age*, September 18, 1978, p. 20.

105 *American Machinist*, April 1977, p. 33.

106 *Ibid.*, October 1977, p. 47.

[107] *Iron Age*, August 21, 1978, p. 14.

[108] *American Metal Market/Metalworking News*, September 11, 1978, p. 31; *Iron Age*, September 11, 1978, p. 16.

[109] *Iron Age*, December 17, 1979.

[110] *Ibid.*, March 17, 1980, pp. 35-46.

[111] *Ibid.*, September 15, 1980.

[112] U.S., Bureau of the Census, Census of Manufactures; *Annual Survey of Manufactures*.

[113] *Iron Age*, March 17, 1980, pp. 35-46.

[114] *Ibid.*, November 23, 1981, p. 95.

[115] *American Machinist*, January 1982, p. 51.

[116] *Annual Reports* - Cincinnati Milacron Inc.

[117] *Iron Age*, February 2, 1981; *American Machinist*, June 1981; January 1982.

[118] *Annual Reports* - Cincinnati Milacron Inc.

[119] *American Metal Market/Metalworking News*, August 25, 1980, pp. 12, 14.

[120] *Iron Age*, November 15, 1979, pp. 83-87.

[121] *Ibid.*, January 21, 1980.

[122] *American Machinist*, September 1977 .

[123] *Automotive Industry*, June 1978, p. 43.

[124] *Iron Age*, February 5, 1979, pp. 28-29; January 21, 1980, May 26, 1980; *American Metal Market/Metalworking News*, January 5, 1981, p. 62; October 5, 1981, p. 14.

[125] Besides the diversification of machine tool builders discussed here, the horizontal integration of the industry was furthered by the entrance of conglomerates into the business. Bendix, Textron, and Litton are now highly placed in machine tool ranks. In mid-1982, a Canadian conglomerate, AMCA, was seeking to take over Giddings & Lewis. (*New York Times*, July 13, 1982, pp. D1, 5.)

[126] *American Metal Market/Metalworking News*, May 18, 1981, pp. 28-29.

[127] *Iron Age*, September 4, 1978, p. 15; *American Metal Market/Metalworking News*, February 1, 1982, p. 12, *New York Times*, March 4, 1982, p. D1.

[128] *American Machinist*, October 1981.

[129] *Annual Reports* - Cincinnati Milacron Inc.; Ex-Cell-O Corporation; Giddings & Lewis, Inc.

[130] U.S., Congress, *Trade with Japan*, August 26, 1980, p. 169.

[131] U.S.. Congress, *Export Policy*, March 20, 1978 (Cross, Gidding & Lewis); *Export Trading Companies Legislation*, July 1, 1980 (Acme-Cleveland); *Annual Reports* - Giddings & Lewis, Inc.

[132] U.S., Congress, *Causes and Consequences of the U.S. Trade Deficit*, November 3, 1977, p. 404.

[133] U.S., Congress, *Export Licensing*, June 14, 1978, p. 28.

[134] *American Machinist*, February 1981.

[135] *American Metal Market/Metalworking News*, June 18, 1979, p. 4; June 25, 1979, p. 6.

[136] U.S., Congress, *Export Licensing*, June 14, 1978, p. 31.

[137] *American Machinist*, May 1978, pp. 43-44.

[138] *Ibid.*, February 1981.

[139] U.S., Bureau of the Census, *U.S. Exports, Schedule B.*

140 U.S., Bureau of Economic Analysis, *Business Statistics*; *Survey of Current Business*.

141 The top ten foreign suppliers to the U.S. in 1981 in descending order were Japan, West Germany, Britain, Taiwan, Switzerland, Canada, Italy, Sweden, France and Spain.

142 U.S., Bureau of the Census, *U.S. Exports: Schedule B*; NMTBA.

143 U.S., Congress, *Machine Tool Programs*, February 7. 1956.

144 U.S., Congress, *Export Policy*, March 20, 1978; *Export Licensing*, June 14, 1978, p. 41; "Statement by Jack E. Moran, Chairman of the Board and Chief Executive Officer, the Carlton Machine Tool Company, before the Joint Economic Committee, September 30, 1981," p. 6 (mimeo).

145 *American Metal Market/Metalworking News*, February 12, 1982, pp. 5-6; May 31, 1982, pp. 1.20.

146 These suits were brought before the International Trade Commission for adjudication under unfair trade practices provisions (Section 337) of U.S. trade law. Consult the following cases: cylinder boring machines (337-32), NC machining centers (337-TA-34), automatic crankpin grinders (337-TA-60), turning machines (337-TA-72), surface grinding machines (337-TA-95).

147 *American Machinist*, February 1981.

148 Houdaille Industries, Inc., *Petition to the President of the United States through the Office of the United States Trade Representative for the Exercise of Presidential Discretion Authorized by Section 103 of the Revenue Act of 1971, 26 U.S.C. 48(a)(7)(D)*, May 3, 1982.

149 Houdaille Industries, Inc., *Petition to the President*, p. 138; also 143-49. See also "Statement by Jack E. Moran," September 30, 1981; Gray, "The American Machine Tool Industry," February 12, 1982.

150 Houdallle Industries, Inc., *Petition to the President*, pp. 145-48.

Chapter V

Export Dependence and International Political Economy

Helping to explain the trade policy of an industry (Chapters III and IV), not to mention entire segments of an economy (Chapter II), is a not insignificant contribution. Yet this study was motivated by the implications that the export dependence thesis holds for national foreign economic policy and international economic relations as well as for theory in international political economy. This brief chapter is devoted to exploring those broader implications.

THEORIES OF PROTECTIONISM

Our findings most directly challenge existing theories of protectionism. Conventional economic explanations were refuted directly in case study and indirectly in aggregate data analyses. The machine tool industry as well as many other industries maintained relatively liberal trade positions despite experiencing poor economic growth, financial performance, labor productivity, technological intensity, and foreign direct investment. International political economy theories of protectionism have also been found to be wanting. Analysts have cited international interdependence, changes in comparative advantage, domestic economic recession, the rise of the welfare state, and the accumulation of surplus industrial capacity as factors exacerbating import penetration and, thereby, pressures for protection. Alternatively, analysts have considered liberal trade relations possible only under the aegis of a hegemonic power. With the relative decline of U.S. dominance over the past two decades and the concomitant decline in hegemonic constraints against protectionism, hegemony theorists envision the growing fragility if not total breakdown of the liberal trade regime.

Where industries face serious import penetration and have experienced severe hegemonic decline without the cross-pressures emanating from export dependence, these political economy explanations hold up. We have, however, paid particular attention to industries cross-pressured by motives for and constraints against protectionism. Despite being seriously penetrated by foreign imports, dependence upon foreign markets for exports has constrained protectionist pressures within this segment of U.S. industry. In addition, the relative international position of the U.S. machine tool industry has deteriorated dramatically and still the industry maintained relatively liberal trade policy views through the 1970s. Among the most important implications of this study is the recognition that import penetration need not be a sufficient condition for protectionism nor hegemony a necessary requisite for liberal trade.

The shortcomings of existing theories of protectionism reflect the general problem of explanations based solely on "immediate causes."[1] Import penetration, hegemonic decline, and poor economic performance all represent motives for protectionism. Yet behavior cannot be explained by motivation alone. One must also have a clear sense of the "permissive conditions" or the constraints on and opportunities for certain outcomes. Export dependence constitutes a significant constraint on protectionism. We have shown that when export dependence levels are high, motivations for protectionism seldom translate into protectionist action. Conversely, protectionism flourishes only when both motives are high and the constraints of export dependence are low.

NORTH-SOUTH AND NORTH-NORTH TRADE RELATIONS

The implications of our findings for theories of protectionism are clear. A subsequent section will go on to discuss further ramifications for other and broader aspects of theory in international political economy. Before doing so, let us examine the implications of the export dependence thesis for trade policy in the broadest of terms. Chapters II-IV examined the empirical impact upon *industrial* trade policy. In this section we consider the implications for *national* foreign economic policy and *international* trade relations.

While the level of export dependence of a particular industry — such as machine tools may experience cyclical fluctuations, aggregate trends in export dependence for the U.S. economy as a whole reveal secular increases. In 1972, 41 industrial sectors representing one-third of U.S. manufacturing shipments exported more than 5% of their production abroad. Seventeen sectors accounting for 13% of the economy sold over 10% of their shipments in foreign markets. Just five years later these figures were 59 sectors representing nearly half of manufactures and 30 sectors constituting 22% of the economy exporting more than 5% and 10%, respectively, of their production abroad (see Table 5.1). In aggregate terms, seven percent of all sales of manufactures were exported in 1979 (see Table 5.2). While this figure is low compared to the levels of other industrial countries, the growth in U.S. export dependence has been impressive. Between 1970 and 1979 and between 1963 and 1979, levels of U.S. export dependence grew 50% and 100%, respectively. This growth rate exceeds changes in the levels of export dependence for such "export-led" countries as West Germany, Japan, Denmark, Norway, Sweden, and others. Only the Netherlands and Ireland have experienced more rapid increases in export dependence.

While the U.S. has been among the most dynamic, all other industrial countries have deepened their export dependence over the past two decades and at levels much higher than that of the U.S. (again, consult Table 5.2). Sweden now exports over one-third of its manufactures abroad, and Austria, Denmark, Ireland, Italy, Norway, and West Germany export approximately one-quarter of their production.

These export dependence trends would seem to bode well for the future of liberal trade relations. Constraints upon protectionism would seem to be even higher in Europe and Japan than in the U.S. where we have already established its impact. Liberal national trade policies would seem to be bolstered on the part of those countries highly dependent upon exports. More importantly, liberal international trade relations in general, may be sustained given that the major industrialized trading partners — those most responsible for the international trade regime — all share high levels of export dependence.

Table 5.1

Share of Manufacturing Shipments Represented by Industrial Sectors According to Level of Export Dependence

Level of Export Dependence	1972	1977
10.00% +	12.6%	21.5%
5.00%-9.99%	25.1%	25.1%
0-4.99%	62.3%	53.4%

Sources: U.S., Bureau of the Census, *Annual Survey of Manufactures: Value of Product_Shipments; U.S. Exports: Domestic Merchandise, SIC-Based Product*, FT 610; *U.S. Exports: SIC Division*, EA 675.

Table 5.2

Export Dependence in Manufactures for Industrial Countries

	Export Dependence (%)			Rate of Change	
	1963	*1970*	*1979*	*1970-79*	*1963-79*
Austria	NA	24.38	25.09 [a]	2.9%	
Canada	9.86	18.93	21.42 [a]	13.2%	117.21%
Denmark	20.91	23.19	25.37 [a]	9.4%	21.3%
France	NA	14.28	19.52	36.7%	
Ireland	7.43	13.57	24.48 [b]	80.4%	229.5%
Italy	NA	21.96	28.32 [a]	29.0%	
Japan	7.50	9.26	11.04 [a]	19.2%	47.2%
Netherlands	5.34	13.37	21.78 [c]	62.9%	307.9%
Norway	20.81	25.56	23.89	6.5%	14.8%
Sweden	20.31	26.00	33.69	29.6%	65.9%
United Kingdom	12.89	15.28	19.18	25.5%	48.8%
United States	3.44	4.61	6.90	49.7%	100.6%
West Germany	17.00	21.73	24.66	13.5%	45.1%

a 1978 figure
b 1976 figure
c 1977 figure

NA not available
Manufactured exports based on SITC 5-8
Gross Output based on ISIC 2-3

Sources: U.N., Statistical Office, *Yearbook of Industrial Statistics; Yearbook of International Trade Statistics; OECD, Statistics of Foreign Trade: Tables by Reporting Countries Series B.*

Yet there is at least one major weakness in these trends — a weakness concerning North-South trade relations.

Two of the most prominent points raised in any discussion of North-South trade relations are quite contradictory politically although empirically compatible. On the one hand, it is often pointed out that less developed countries will increasingly serve as vital and vibrant markets for goods from industrial countries. This market value would seem to aid the integration of LDCs into the world economy. On the other hand, such integration may be impeded by the problems posed by LDC exports (in contrast to their markets). The disruptiveness of LDC exports in the markets of industrial economies has spurred significant attempts to regulate and limit the participation of developing countries in world markets; i.e., trade protection in the form of orderly marketing arrangements, voluntary export restraints, and other informal and discriminatory restraints.

Light can be shed on this dilemma of market value vs. export disruptiveness and enhanced vs. constrained LDC integration into the world economy by analysis based upon export dependence.

To begin with, industrial countries are not heavily dependent upon markets in developing countries (see Table 5.3). While overall export dependence levels for developed countries have grown and are significant, the economies of North America and Western Europe have grown largely dependent upon one another rather than with developing countries. In only one case (Japan) are levels of export dependence upon LDC comparable to dependence upon developed countries. While export dependence on industrial countries ranges from 4% (U.S.) to 27% (Sweden), export dependence on LDC markets ranges from 1% (Ireland) to only 7% (Italy). Levels of export dependence within the North exceeds levels of Northern dependence on the South by one and a half times for the U.S. and up to fifteen times for Ireland.

Not only are levels of industrial country export dependence on LDCs lower in absolute terms than dependence on industrialized markets, but dependence within the North has grown faster than dependence on the South. In seven out of eight cases, rates of growth in intra-North dependence exceed that of North-South dependence between 1963 and 1979 (Japan being the only exception).

Table 5.3

Industrial Country Export Dependence on Other Industrial Countries versus Developing Countries

	EXPORT DEPENDENCE ON:							
	Industrial Countries				Developing Countries			
	level (%)			rate of change	level (%)			rate of change
	1963	*1970*	*1979*	*1963-79*	*1963*	*1970*	*1979*	*1963-79*
Austria	NA	17.67	19.32[a]		NA	0	1.76[a]	
Canada	7.96	16.92	18.95[a]	138%	1.88	1.95	2.27[a]	21%
Denmark	15.03	18.82	19.77[a]	32%	4.29	3.31	4.60[a]	7%
France	NA	14.28	19.52		NA	3.59	5.21	
Ireland	6.86	12.83	22.77[b]		0	0	1.49[b]	
Italy	NA	15.94	19.58[a]		NA	4.63	7.39[a]	
Japan	3.51	4.57	5.03[a]	43%	3.96	4.18	5.59[a]	41%
Netherlands	4.30	11.28	17.99[c]	318%	0.94	1.78	3.22[c]	243%

(continued)

Table 5.3 (continued)

	EXPORT DEPENDENCE ON:							
	Industrial Countries				Developing Countries			
	level (%)			rate of change	level (%)			rate of change
	1963	*1970*	*1979*	*1963-79*	*1963*	*1970*	*1979*	*1963-79*
Norway	15.15	22.30	18.48	22%	5.25	2.46	4.80	-9%
Sweden	11.99	20.86	26.74	123%	7.47	3.64	5.53	-26%
United Kingdom	6.43	9.25	12.94	101%	6.04	5.42	5.79	-4%
United States	1.69	2.90	4.04	139%	1.74	1.69	2.81	62%
West Germany	12.98	17.08	18.61	44%	3.50	3.72	4.73	35%

a 1978 figure NA not available

b 1976 figure Manufactured exports based on SITC 5-8.

c 1977 figure Gross Output based on ISIC 2-3.

Sources: U.N., Statistical Office, *Yearbook of Industrial Statistics; Yearbook of International Trade Statistics; OECD, Statistics of Foreign Trade: Tables by Reporting Countries, Series B.*

For the U.S., dependence upon markets in Western Europe, Japan, and Canada deepened 139% between 1963 and 1979, while dependence upon developing countries grew only 62% over the same period. In 1977, only 16 U.S. industrial sectors exported over 5% of their production to LDC markets in contrast to 32 sectors exporting at such levels to developed markets (see Table 5.4). Nine sectors shipped over 10% of their output to industrial countries while only four did so to developing countries.

While the dependence of the major industrial trading nations upon LDCs remains low in absolute terms and has grown at rates below that of overall export dependence, increasing proportions of manufactures are nevertheless being shipped to these developing markets. It is of utmost importance, however, to be specific about which sectors have been making increasing shipments to LDCs and such levels of specificity can be provided for the U.S. case.

Table 5.5 lists the fifteen U.S. industrial sectors with the highest degrees of dependence upon LDC markets. In the case of aircraft (SIC 372), 19% of domestic production is shipped to developing countries. In line with our thesis, we would hypothesize that these sectors would work to foster liberal trade relations between the U.S. and LDCs.

Yet we have found in our analysis in Chapters II-IV that export dependence plays its greatest role under adversity. That is, the importance of sectors being dependent upon exports is heightened when those sectors may be motivated to seek trade protection. Constraints on protectionism are less important for industries less inclined to go down the protectionist road than for those sectors that may be tempted to do so. In short, in proposing one of this study's few policy-type recommendations, one would ideally want high levels of export dependence to "absorb" protectionist pressures in sectors threatened by foreign imports.

Such absorption of import penetration is *not* taking place in U.S.-LDC trade relations. In reference to the top 15 sectors dependent upon LDC markets, only one sector (representing only 9% of the sales of the 15 sectors) is also penetrated by imports (SIC 367, electronic components). In essence, what little U.S. export dependence there is on LDC markets is "wasted;" little is offered in directly offsetting the very real disruptiveness of imports from LDCs.

Table 5.4

U.S. Industrial Sectors Most Highly Dependent upon Exports to Industrial Countries versus Developing Countries, 1977

Sectors Dependent upon Industrial Countries (SIC Code & Title)	*Sectors Dependent upon Developing Countries (SIC Code & Title)*
10.00%+ Export Dependence	10.00%+ Export Dependence
241. Logging	299. Misc. Petroleum Products
261. Pulp Mills	353. Construction Machinery
299. Misc. Petroleum Products	356. General Industrial Machinery
353. Construction Machinery	372. Aircraft
355. Special Industrial Machinery	
357. Office & Computing Machines	
372. Aircraft	
382. Measuring Devices	
386. Photographic Equipment	
5.00-9.99% Export Dependence	5.00-9.99% Export Dependence
207. Fats & Oils	207. Fats & Oils
237. Fur Goods	211. Cigarettes
283. Drugs	221. Cotton Weaving Mills
286. Industrial Organic Chemicals	287. Agricultural Chemicals
287. Agricultural Chemicals	289. Misc. Chemical Products
289. Misc. Chemical Products	351. Turbines
311. Leather Tanning	355. Special Industrial Machinery
321. Flat Glass	358. Refrigeration Machinery
323. Glass Products	367. Electronic Components
346. Metal Forgins	374. Railroad Equipment
351. Turbines	382. Measuring Devices
352. Farm Machinery	387. Watches
354. Metalworking Machinery	261. Pulp Mills
356. General Industrial Machinery	

(continued)

Table 5.4 (continued)

Sectors Dependent upon Industrial Countries (SIC Code & Title)	*Sectors Dependent upon Developing Countries (SIC Code & Title)*
5.00-9.99% Export Dependence (continued) 358. Refrigeration Machinery 367. Electronic Components 369. Misc. Electrical Equipment 371. Motor Vehicles 383. Optical Instruments 384. Medical Instruments 391. Jewelry, Sliverware 393. Musical Instruments 395. Pens, Pencils, Office Supplies 281. Industrial Inorganic Chemicals	5.00-9.99% Export Dependence (continued)

Source: see Table 5.1

Table 5.5

Top 15 U.S. Industrial Sectors Dependent
upon Exports to Developing Countries

SIC Code & Title	Export Dependence Ratio (%)*	Import Penetration Ratio(%)*	"High" §
1 372 Aircraft	19.03	0.18	
2 353 Construction Machinery	15.98	0.17	
3 299 Misc. Petroleum Prods	13.44	0	
4 356 General Indstrl Machy	9.66	0.40	
5 367 Electronic Components	9.52	9.27	X
6 355 Special Indstrl Machy	8.64	0.50	
7 358 Refrigeration Machy	8.42	0.23	
8 207 Fats & Oils	8.24	3.42	
9 289 Misc. Chemical Prods	8.19	1.56	
10 287 Agricultural Chemicals	7.69	0.55	
11 351 Turbines	6.63	0.41	
12 382 Measuring Devices	6.62	0.59	
13 261 Pulp Mills	6.22	1.67	
14 211 Cigarettes	6.12	0	
15 374 Railroad Equipment	6.04	0.36	
COMPOSITE	10.73	1.56	1

* regarding trade with developing countries only
§ above 5.25%—weighted mean for entire data set

Sources: see Table 5.1

The situation is quite (and fortunately) different for U.S.-developed country trade relations. Table 5.6 lists the top 15 sectors which are most highly dependent upon sales to industrialized markets. In stark contrast to U.S.-LDC relations, this second set of sectors absorb a good deal of import penetration and, therefore, protectionist pressure. Eight of the 15 highly dependent sectors (representing 48%

of the sales to industrial markets) absorb four times the import penetration of sectors dependent upon LDC markets (6.3% vs. 1.6%). Export dependence in the one setting plays an important role in constraining protectionism while export dependence in the other case faces little protectionist pressure to constrain.

This analysis can be turned on its head by considering the sources of import penetration problems. Just as one would want highly export dependent sectors to "absorb" pressures for protection, one would also want sectors penetrated by imports to have some stake in export markets. We would want sectors with significant protectionist incentives to be "cross-pressured" by constraints against protectionism.

Such cross-pressuring is taking place in U.S. trade relations with other industrial countries. Table 5.7 lists the fifteen sectors most penetrated by imports from industrial suppliers. Leading the list is motorcycles (SIC 375) for which industrial suppliers (largely Japan) are responsible for capturing nearly 50% of the U.S. market. As indicated, six of these sectors accounting for 78% of the "top l5's" sales are also relatively highly export dependent.

In contrast once again, those sectors threatened by imports from LDCs are rarely cross-pressured (see Table 5.8). Only two of the fifteen sectors penetrated by LDC exports are also relatively highly dependent upon developing markets. Taken as a whole, sectors penetrated by LDC suppliers are cross-pressured by one-half the export dependence of their counterparts threatened by industrial country goods (2.8% vs. 5.7%). While export dependence may work to constrain U.S. protectionism against its industrial trading partners, lower constraints exist for U.S. action against LDC producers.

Indeed, this possibility can be confirmed by examining the protectionist behavior of these sectors penetrated either by industrial or developing countries. Table 5.9 compares the protectionist activity of those highly penetrated sectors listed in Tables 5.7 and 5.8 for 1975 through 1979. Clearly those sectors penetrated by LDCs and less cross-pressured by dependence upon LDC markets exhibit significantly greater levels of protectionist activity. Ten of the 15 sectors initiated 55 trade actions involving approximately 62% each of the sectors' shipments and imports. In comparison, sectors highly penetrated by imports from industrial but cross-pressured by exports to those markets exhibited much lower levels of protectionist activity. Only seven of the 15 sectors took 15 trade actions involving 44% of the sectors' imports and only 16% of their shipments.

Table 5.6

Top 15 U.S. Industrial Sectors Dependent upon Exports to Industrial Countries, 1977

SIC Code & Title			Export Dependence Ratio (%)*	Import Penetration Ratio(%)*	"High" §
1	299	Misc. Petroleum prods	30.69	3.08	
2	261	Pulp Mills	18.21	31.47	X
3	357	Office & computing equip	17.07	6.81	X
4	241	Logging	16.33	0.87	
5	382	Measuring Devices	14.12	3.86	
6	372	Aircraft	11.13	3.84	
7	353	Construction Machinery	11.11	3.76	
8	386	Photographic Equip	10.76	0.83	
9	355	Special Indstrl Machy	10.44	13.53	X
10	384	Medical Instruments	9.45	2.81	
11	352	Farm Machinery	9.16	8.18	X
12	351	Turbines	8.39	6.31	X
13	393	Musical Instruments	8.39	10.89	X
14	281	Indstrl Inorganic Chems	8.31	6.14	X
15	356	General Indstrl Machy	8.05	5.57	X
COMPOSITE			11.00	6.29	8

* regarding trade with industrial countries only
§ above 5.25% -- weighted mean for entire data set

Sources: see Table 5.1

Table 5.7

Top 15 U.S. Industrial Sectors Penetrated by Imports from Industrial Countries, 1977

SIC Code & Title			Import Penetration Ratio (%)*	Export Dependence Ratio(%)*	"High" §
1	375	Motorcycles	47.84	1.30	
2	261	Pulp Mills	31.47	18.21	X
3	365	Radio & Television	25.45	3.26	
4	326	Pottery	23.23	2.33	
5	387	Watches	19.24	2.83	
6	383	Optical Instruments	17.84	6.24	X
7	242	Sawmills	15.80	4.60	
8	385	Opthalmic goods	15.23	3.71	
9	391	Jewelry, Silverware	14.93	6.62	X
10	302	Rubber Footwear	14.80	0.33	
11	262	Paper Mills	14.34	1.61	
12	355	Special Industrial Machy	13.53	10.44	X
13	314	Nonrubber Footwear	12.70	0.44	
14	371	Motor Vehicles	12.21	6.05	X
15	393	Musical Instruments	10.89	8.39	X
COMPOSITE			14.57	5.66	6

* regarding trade with industrial countries only
§ above 5.25% — weighted mean for entire data set

Sources: see Table 5.1

Table 5.8

Top 15 U.S. Industrial Sectors Penetrated by
Imports from Developing Countries, 1977

SIC Code & Title			Import Penetration Ratio (%)*	Export Dependence Ratio(%)*	"High" §
1	302	Rubber Footwear	46.45	0.28	
2	238	Misc. Apparel	24.41	0.80	
3	315	Leather Gloves	18.80	4.03	
4	317	Leather Handbags	18.37	0.43	
5	391	Jewelry, Silverware	15.89	3.89	
6	387	Watches	15.62	5.18	X
7	316	Luggage	15.04	3.94	
8	236	Misc. Outwear	14.11	1.08	
9	365	Radio & Television	12.26	2.49	
10	206	Sugar	11.83	0.29	
11	394	Toys & Sporting goods	10.79	1.88	
12	396	Costume Jewelry	10.39	1.48	
13	314	Nonrubber Footwear	10.32	0.28	
14	333	Primary nonferrous metals	9.82	0.19	
15	367	Electronic Components	9.27	9.52	X
COMPOSITE			12.12	2.84	2

* regarding trade with developing countries only
§ above 5.25% — weighted mean for entire data set

Sources: see Table 5.1

Table 5.9

Protectionist Actions Undertaken by U.S. Industrial Sectors
Most Highly Penetrated by Imports, 1975-1979

Top 15 Sectors Penetrated by Imports from Industrial Countries			
		% Top 15 Shipments & Imports Involved	
SIC Code	Protectionist Action*	*S*	*M*
375	none		
261	none		
365	77EC Televisions - Japan 78EC CB Radios - Japan	0.3	7.9
326	78EC Earthenware tableware articles - Japan - Britain	0.7	1.3

Top 15 Sectors Penetrated by Imports from Developing Countries			
		% Top 15 Shipments & Imports Involved	
SIC Code	Protectionist Action*	*S*	*M*
302	76CVD Rubber footwear - South Korea - Taiwan (2)	0.6	4.6
238	76EC Cotton gloves - Hong Kong - Taiwan - South Korea 77CVD Leather wearing apparel - Taiwan - South Korea	10.9	0.3

(continued)

Table 5.9 (continued)

Top 15 Sectors Penetrated by Imports from Industrial Countries				Top 15 Sectors Penetrated by Imports from Developing Countries			
		% Top 15 Shipments & Imports Involved				% Top 15 Shipments & Imports Involved	
SIC Code	Protectionist Action*	*S*	*M*	SIC Code	Protectionist Action*	*S*	*M*
387	none			238 (cont)	78CVD Textiles - Uruguay - Philippines 78CVD Leather wearing apparel - Uruguay (2) 79CVD Leather wearing apparel - Argentina - Colombia - Brazil		
383	none						
242	none						
385	none						
391	76EC Stainless steel flatware - Japan 78EC Stainless steel flatware - Japan	1.7	2.2				
302	none						
262	78AD Offset paper - Canada 79AD Kraft condenser paper - France	7.2	6.9	315	76EC Leather Gloves - Hong Kong - Taiwan - South Korea	0.2	0.4

(continued)

Table 5.9 (continued)

Top 15 Sectors Penetrated by Imports from Industrial Countries			
		% Top 15 Shipments & Imports Involved	
SIC Code	Protectionist Action*	*S*	*M*
355	none		
314	76EC Nonrubber footwear - Italy 77EC Nonrubber footwear - Italy	2.1	2.0
371	none		
393	78CVD Musical instruments - Japan	0.5	0.3

Top 15 Sectors Penetrated by Imports from Developing Countries			
		% Top 15 Shipments & Imports Involved	
SIC Code	Protectionist Action*	*S*	*M*
317	76CVD Leather handbags - Brazil 77CVD Leather handbags - Taiwan - South Korea 78CVD Leather handbags - Colombia - Uruguay	0.6	4.6
391	76EC Stainless steel flatware -South Korea -Taiwan	4.5	6.9

(continued)

Table 5.9 (continued)

Top 15 Sectors Penetrated by Imports from Industrial Countries				Top 15 Sectors Penetrated by Imports from Developing Countries			
		% Top 15 Shipments & Imports Involved				% Top 15 Shipments & Imports Involved	
SIC Code	Protectionist Action*	*S*	*M*	SIC Code	Protectionist Action*	*S*	*M*
				391 (cont)	78EC Stainless steel flatware -South Korea -Taiwan -Hong Kong		
				387	none		
				316	none		
				236	78CVD Textiles -Argentina -Taiwan	8.3	9.4

(continued)

Table 5.9 (continued)

Top 15 Sectors Penetrated by Imports from Industrial Countries				Top 15 Sectors Penetrated by Imports from Developing Countries			
		% Top 15 Shipments & Imports Involved				% Top 15 Shipments & Imports Involved	
SIC Code	Protectionist Action*	*S*	*M*	SIC Code	Protectionist Action*	*S*	*M*
				365	77EC Televisions - Taiwan - South Korea 78EC CB Radios - Taiwan - South Korea	8.6	11.0
				206	77EC Sugar - Dominican Republic - Philippines - Peru - Guatemala	15.7	14.0

(continued)

Table 5.9 (continued)

Top 15 Sectors Penetrated by Imports from Industrial Countries				Top 15 Sectors Penetrated by Imports from Developing Countries			
		% Top 15 Shipments & Imports Involved				% Top 15 Shipments & Imports Involved	
SIC Code	Protectionist Action*	*S*	*M*	SIC Code	Protectionist Action*	*S*	*M*
				394	78EC Fishing tackle - Taiwan - Mexico - South Korea	7.4	6.3
				396	none		
				314	76EC Nonrubber footwear - Taiwan - South Korea 77EC Nonrubber footwear - Taiwan - South Korea 76CVD Nonrubber footwear - South Korea - Argentina	5.5	5.2

(continued)

Table 5.9 (continued)

Top 15 Sectors Penetrated by Imports from Industrial Countries				Top 15 Sectors Penetrated by Imports from Developing Countries			
		% Top 15 Shipments & Imports Involved				% Top 15 Shipments & Imports Involved	
SIC Code	Protectionist Action*	*S*	*M*	SIC Code	Protectionist Action*	*S*	*M*
				314 (cont)	78CVD Nonrubber footwear - Uruguay - India 79CVD Nonrubber footwear - Argentina - India		
				333	none		
				367	none		
	TOTAL	16.2	43.6		TOTAL	62.9	62.3

(continued)

* "Protectionist Action" cites year action taken; type of action (EC=escape clause, AD=antidumping, CVD=countervailing duty); product involved; and countries targeted

Table 5.9 (continued

Sources: U.S. Department of the Treasury, *Customs Bulletin; Customs Bulletin & Decisions*; U.S., *Federal Register; U.S., International Trade Commission, Annual Report; Operation of the Trade Agreements Program; USITC Publications*; U.S., Office of the United States Trade Representative, *Trade Action Monitoring System*; U.S., President, *Annual Report of the President of the United States on the Trade Agreements Program*

Much anxiety has been expressed over the "non-complementarity" of U.S. trade relations with other industrial countries.[2] Industrial countries trade not in complementary but in similar products given the similarity of their industrial structures and level of economic development. Such trade poses immediate and direct problems of competition and industrial conflict.

North-South trade has been more complementary. Countries at markedly different levels of industrial development tend to specialize in different products. Exchange between North and South is supposed to be less directly competitive and conflictual. Whether or not these propositions are correct, this study reveals an important benefit of "non-complementarity" in trade and economic structure. By trading in similar products, as industrial countries do, there is a greater chance that import penetrated sectors will also be export dependent. True, pressures for protection may be heightened by direct industrial competition, but constraints on protectionism would also be in force. In contrast, where trade is more complementary, exports from developing countries may be disruptive in industrial markets without the besieged industries having any stake in LDC markets. If ideal trade relations of low motivation for and high constraint against protectionism are unrealistic, then the second best solution of high motivation and high constraint (North-North trade) is to be preferred to a situation of high motivation and low constraint (North-South trade).

In summary, industrial countries are much more dependent upon each other's export markets than on LDC markets. In addition, U.S. sectors dependent upon industrial countries "absorb" more protectionist pressures than sectors dependent upon developing countries. And sectors threatened by imports from industrial countries are more "cross-pressured" by export dependence than their counterparts threatened by LDC suppliers. The other industrial countries most likely share these characteristics of U.S.-industrial country and U.S.-LDC trade relations. These findings prove ominous for developing countries seeking participation in the world economy. Overall, what may be as significant constraint on protectionism among industrial countries may not exist for industrial country commercial policy towards LDCs.

NATIONAL POWER AND THE STRUCTURE OF THE INTERNATIONAL POLITICAL ECONOMY

In this section we not only consider the ramifications of this study for broad theory in international political economy, but expand our concern from export dependence to *general* international *economic dependence*. Drawn into discussion here is dependence upon imports, capital, finance, and technology. The implications of general dependence, and especially mutual economic dependence, are drawn out for conceptualizations of national power and alternative structures of the world political economy.

National economic dependence has long been intimately linked with national power.[3] Emerson has proposed that "the power of A over B is equal to and based upon the dependence of B upon A."[4] Blau writes that the source of power is "one-sided dependence."[5] The purpose of Hirschman's classic study, *National Power and the Structure of Foreign Trade*, was to demonstrate economic dependence as a "root cause" of international influence.[6] Keohane and Nye have cited "asymmetrical interdependencies as sources of power among actors."[7] Caporaso reiterates, "power springs from asymmetrical interdependence"[8] and Baldwin observes, "dependency relations are a form of influence relations."[9] Waltz presents the obverse of these points by arguing that minimal economic dependence offers freedom from external influence.[10]

No one could argue with the proposition that power advantages emanate from highly asymmetrical relations of economic dependence. Such highly asymmetrical relations serve as the ideal type for the above observations linking dependence and power. Blau's analysis is based on "one-sided" or "unilateral" dependence. In discussing how best to study the contribution of international economic relations to national power, Hirschman concludes that "colonial trade relations often gives us the opportunity of observing this type of cumulative effect" and goes on to use Nazi domination over Eastern Europe as his case study.[11]

But what are the power implications in situations where dependence relations are less asymmetrical? While thoroughly "symmetrical" or "balanced" relations are thoroughly idealistic, conditions of "two-way," "mutual," or "shared" dependence — while always technically asymmetrical — should be distinguished from "highly" or "one-way" asymmetrical relations. Consideration of the status of national power under conditions of mutual economic dependence takes on all the greater significance since mutual dependence may characterize relations among industrial countries as proposed in the previous section.

National power (or policy autonomy) may be limited by dependence in two ways. The first is through direct pressure or coercion from external entities upon which a nation may be dependent.

Such constraints are based on the behavioral power of others that grows from relations of dependence.

The second source of constraint is the recognition and internalization of a disadvantageous position within structured power relations. When faced with the structural power of others, constraints need not be imposed from outside but conceded from within. In our own empirical analysis, we demonstrated how industrial trade policy is constrained by export dependence. The link to national power is based on the working assumption that industrial opposition to certain policy options heightens constraints on national policy. Such restraints need not be completely effective, though national policy autonomy would certainly be enhanced in the absence of domestic resistance.

When conditions of mutual dependence are operative, it follows that internalized constraints may be in effect in a number of countries simultaneously. National power on the part of all concerned may be limited. Most importantly, when nations are dependent upon one another, the manipulation of such dependence relations for power purposes may be circumscribed. Unlike conditions of mutual sensitivity interdependence, shared dependence can be manipulated only at great cost. When two countries can "affect" each other only in relatively insignificant (however extensive) ways, each could manipulate or threaten to manipulate their mutual sensitivity with little domestic cost and resistance.

In contrast, where two countries are more than just mutually sensitive and are, instead, heavily reliant upon one another economically, the manipulation of that dependence, for either country, will face great resistance domestically. It makes little difference to a domestic interest that the "net" result of a power move may favor "us" over "them." Domestic interests will, instead, be centrally concerned with their own losses. If those losses are deemed unacceptable, then domestic interests will care little whether foreign losses are greater or not. Governments may recognize that foreign dependence may be greater than domestic dependence, but efforts to reap advantages from such asymmetry will be hindered by domestic resistance.

In sum, the postulated link between power and dependence must be substantially qualified and revised, at least with respect to the all-important cases of mutual dependence. It is not impossible, of course, to manipulate relations of mutual dependence, but it will be much more costly to do so. The domestic ramifications of dependence may effectively inhibit the exercise of international power among countries that are dependent upon each other.

Sir Norman Angell provides a stunning analogy for this argument in reference to two men in a boat:

> The boat was leaky, the sea heavy, and the shore a long way off. It took all the efforts of the one man to row, and of the other to bail. If either had ceased both would have been drowned. At one point the rower threatened the bailer that if he did not bail with more energy he would throw him overboard; to which the bailer made the obvious reply that if he did, he (the rower) would certainly drown also. And as the rower was really dependent upon the bailer, and the bailer upon the rower, *neither could use force against the other*.[12]

I would only add that even when the relationship is not strictly symmetrical "reciprocal dependency constrains each party's behavior with respect to each other."[13]

Two major implications flow from this discussion with regard to how we conceptualize economic dependence and how we view international relations generally. The first concerns the *relativity* of dependence. Power analysts have overcome the notion of power as an attribute by successfully arguing that power can only be understood within the context of a social relationship. In Emerson's words, "to say 'X has power' is vacant, unless we specify 'over whom'."[14] Sensitivity interdependence is also relative; some countries may be more sensitive to external intervention than others.

And analysts argue that economic dependence is also relative. Blau emphasizes that particular attention should be given to "degrees" of dependence.[15] Baldwin decries the "lump-of-dependency fallacy;" dichotomizing countries as either dependent or not without reference to degree.[16] Caporaso makes the definitive statement: "Dependence as asymmetric interdependence is immediately a dyadic concept and a "net" concept, i.e., it is measured by looking at the differential between A's reliance on B and B's reliance on A."[17]

There is no doubt that dependence is empirically relative. Indeed, this study has been very careful to measure and take note of differing levels of dependence across sectors and countries. The major purpose in doing so was to examine the extent of policy constraint. In contrast, the major motivation behind Emerson and company's concern for the relativity of dependence is to shed light on power differentials. Yet in light of the preceding discussion, serious questions are raised as to the validity of inferring international leverage from levels of dependence. In particular we have argued that differentials in dependence do not yield power advantages, not at least under conditions of mutual and relatively high levels of dependence. If the bilateral relations of two countries are constrained by mutual dependence, their varying levels of dependence, while empirically measurable, are politically irrelevant (at least internationally). As far as power is concerned, the fact that one country may be more dependent than another means little when the other is also dependent on the one.

The second implication to emerge from this discussion concerns the *zero-sum* or *non-zero-sum* characteristics of international economic dependence. Since power is relative, the power game is zero-sum: more for A means less for B and at no time can all gain at once.

In stark contrast, dependence clearly has non-zero-sum characteristics. Empirically, countries can and have grown more dependent together. Conceptually, increasing dependence for one country does not reduce the dependence of another.

If these propositions have any validity and if countries, especially the industrial trading partners, continue to grow more dependent upon each other, then longstanding disaffection for traditional conceptions of international relations should be magnified. If mutual dependence comes to increasingly characterize international economic relations, then the disjunction between dependence and power and the effective non-relativity of dependence will diminish or remove one major "chessboard" from being subject to international competition.[18] While the objective of Hirschman's classic study was to highlight the "extensive use of international relations as an instrument of national power,"[19] increasing mutual dependence may shackle such policy instruments. In a world that has placed a premium on relative position, mutual dependence resists being subject to such concerns.

Similarly, the non-zero-sum characteristics that mutual dependence may confer upon international economic relations may rival the constant-sum game of strategic-military affairs. While even relative weaknesses may yield advantages for one country over another in the political-military "world," countries can grow more dependent and constraints on their policies can mount without creating power advantages for others in an economic "world" of mutual dependence.[20] To use Angell's analogy, all nations can be in the same "boat" of dependence.

We close this section with a less abstract but still general discussion of the ramifications of this study for the structure of the international political economy. We have already criticized hegemonic stability theory as an explanation of protectionism. We can expand that criticism to the theory's view of alternative structures of global economic relations.

According to the argument,[21] there are only two major alternative structures for the international political economy. One is a hegemonic system characterized by unequal distribution of international power, asymmetrical relations of dependence, and open and liberal international economic relations. The beneficiary of the unequal distribution of power, the hegemon, has both the resources and incentive to construct a liberal international economic regime. Asymmetrical dependence between hegemonic and non-hegemonic powers is both enhanced by and provides leverage for the hegemon's structuring of a liberal regime.

The theory's other alternative is an autarkic system of more equal distribution of power, low levels of cross-national dependence, and closed and protectionist international economic relations. Without the "integrating" force of a hegemonic power, a liberal regime breaks down. Low levels of external dependence makes affordable and are enhanced by the severing of international economic links. The theory of hegemonic stability offers either one view of the world or the other. Any intermediate condition is either transitory or "fragile" and always short-lived.

Yet this study offers a glimpse at a third vision for the international political economy. Between the asymmetrical dependence of a hegemonic system and economic autarky stands economic relations of high and mutual dependence. From our finding that export dependence can constrain industrial trade policy, we have extrapolated that economic dependence can constrain national foreign economic policy. Furthermore, if all or major countries become mutually dependent, then international economic relations in general may be moderated.

Mutual economic dependence may work to constrain protectionism in an international system just as hegemony may. More importantly, mutual economic dependence may have this effect *irrespective* of the absence or presence of hegemony. Just as we have concluded that liberal industrial trade policy in the midst of industrial hegemonic decline refutes theories of protectionism based upon hegemony, the possibility of liberal international economic relations in the absence of a hegemonic power modifies hegemonic stability theory's visions of the alternatives for international political economy. Given the independent impact that mutual economic dependence may play, hegemony is reduced from a necessary to only a sufficient condition for liberal trade relations.

With the decline of U.S. economic and political hegemony, the question arises as to the fate of the liberal international economic regime, For rigorous hegemony stability theorists there is but one, unfortunate answer: progressive deterioration. If countries, especially the major industrialized trading partners, are in fact growing mutually dependent, then this study holds out another possibility: the sustenance of the liberal regime. We may be moving toward a historically unique international political economy characterized by more symmetrical distribution of international power, mutual dependence, and liberal — not protectionist — international economic relations.

CONCLUSION

We have conducted a rather labyrinthian analysis and explication of the link between export dependence and trade policy at a number of different levels of analysis. At the most disaggregated level, we closely examined a single crucial case, the U.S. machine tool industry (Chapters III and IV). The illustrative case was "crucial" because the export dependence thesis was posed against a number of other prominent explanations that would seem to apply to the machine tool industry. The competing explanations were cast aside given the fact that the industry maintained relatively liberal trade policy views through the 1970s — a decade in which the industry suffered severe import penetration, hegemonic decline, poor economic performance, deteriorating technological advantage, and low foreign direct investment. In explaining this anomaly, we found that high levels of export dependence effectively constrained protectionist pressures within the industry. We also found that when export dependence deteriorated in the 1950s and 1960s, the door was opened for the ascendance of protectionist sentiments. Such a deterioration once again took place in the 1980s.

At a higher level of analysis, we applied the export dependence thesis across 129 U.S. industrial sectors (Chapter II). We found that the machine tool industry was part of an entire segment of U.S. industry that not only manifested little protectionist activity but did so in spite of high levels of import penetration. In the aggregate data analysis, high levels of export dependence were, again, directly related to the low levels of protectionism exhibited by this anomalous yet significant portion of the economy. Low levels of export dependence were found to permit other segments of industry to go down the protectionist road.

Chapter I cast the the export dependence thesis at the national and international levels as well as industrial level. In the present chapter, we explored such broader applications. The thesis was first extended to national foreign economic policy. It was found that while constraints of export dependence may be increasing in the aggregate for the U.S. and other industrial countries, such restraints may be less effective in developed country trade relations with LDCs than in relations among industrial countries.

At yet a higher level of analysis, implications were drawn out for international economic relations. The constraints of export dependence may apply at the international as well as the national level as most of the major economic powers become more mutually dependent. Mutual dependence not only can underlie the sustenance of the liberal international economic regime in the absence of a hegemonic power, but can challenge traditional conceptions of international relations generally. The rise of mutual economic dependence can open the door to a world of fewer opportunities for the exercise of power, less concern about relative position, and reduced fear of vulnerability to manipulation.

This study has sought to chart and explain the strengthening of two rather idealistic ends: liberal commercial policy and liberal international trade relations. Yet our conclusions have not been established by idealistic means. The export dependence thesis rests neither on the propagation of free trade ideology,[22] nor on some apocalyptic vision that spurs international cooperation,[23] nor on voluntaristic political will or leadership.[24] Neither has our argument been based on more subtle cognitive developments among managers of international regimes.[25]

Instead, the study has rested upon realistic analysis of concrete imperatives working to maintain liberal trade. Barring any alteration in human nature, any meaningful response to the turbulence of world developments demands realistic, intensive scrutiny but without losing sight of preferred ideals.

NOTES

1 Again, consult Kenneth N. Waltz, *Man, the State and War: A Theoretical Analysis* (New York: Columbia University Press, 1959), pp. 232-36 for a discussion of the terms "immediate causes" and "permissive conditions."

2 U.S., Congress, Senate, Committee on Finance, Subcommittee on International Trade, *U.S. International Trade Strategy*, hearing, July 28, 1980, 96th Congress, Second Session.

3 Such links have, of course, been established for social entities other than nation-states.

4 Richard M. Emerson, "Power-Dependence Relations," *American Sociological Review*, vs. 27, n. 1 (February 1962), p. 32.

5 Peter M. Blau, *Exchange and Power in Social Life* (New York: Wiley, 1964), p. 118.

6 Albert O. Hirschman, *National Power and the Structure of Foreign Trade* (Berkeley: University of California Press, 1945), p. 17.

7 Robert O. Keohane and Joseph S. Nye, *Power and Interdependence: World Politics in Transition* (Boston: Little, Brown, 1977), p. 18.

8 James A. Caporaso, "Dependence, Dependency, and Power in the Global System: A Structural and Behavioral analysis," *International Organization*, v. 32, n. 1 (Winter 1978), p. 28.

9 David A. Baldwin, "Interdependence and Power: A Conceptual Analysis," *International Organization*, v. 34, n. 4 (Autumn 1980), p. 501-502.

10 Kenneth N. Waltz, *Theory of International Politics* (Reading, Mass.: Addison-Wesley, 1979), pp. 138-160. See also Waltz, "The Myth of National Interdependence" in *The International Corporation*, Charles P. Kindleberger, ed. (Cambridge: MIT Press, 1970).

11 Hirschman, *National Power*, p. 13.

12 Sir Norman Angell, *The Foundations of International Polity* (London:

William Heinemann, 19l4), p. 17, cited in Baldwin, "Interdependence and Power," p. 483, emphasis added.

[13] Baldwin, "Interdependence and Power," p. 484.

[14] Emerson, "Power-Dependence Relations," p. 32.

[15] Blau, *Exchange and Power*, pp. 119-20.

[16] Baldwin, "Interdependence and Power," p. 448.

[17] Caporaso, "Dependence, Dependency, and Power," p. 18.

[18] The chessboard analogy is Hoffman's. Stanley Hoffman, "Notes on the Elusiveness of Modern Power," *International Journal*, 30 (Spring 1975).

[19] Hirschman, *National Power*, p. xv.

[20] The reference to "worlds" is derived from Richard Rosecrance, "The One World of Hans Morgenthau," *Social Research*, v. 48, n. 4 (Winter 1981).

[21] See Chapter I references.

[22] Judith L. Goldstein, [citation from Bibiography].

[23] Richard A. Falk, *A Study of Future Worlds* (New York: Free Press, 1975).

[24] Interestingly, hegemony stability theorists often uncharacteristically resort to anti-structuralist pleas for political voluntarism to avoid the economic nationalism they are led to predict given the decline in U.S. hegemony. See David P. Calleo and Benjamin M. Rowland, *America and the World Political Economy: Atlantic Dreams and National Realities* (Bloomington, Indiana University Press: 1973), pp. 258-59; Robert Gilpin, *U.S. Power and the Multinational Corporation: The Political Economy of Foreign Direct Investment* (New York: Basic Books, 1975), p. 262.

[25] Oran R. Young, "International Regimes: Problems of Concept Formation" and Ernst B. Haas, "Why Collaborate? Issue-Linkage and International Regimes," both in *World Politics*, v. 32, n. 3 (April 1980); *International Regimes*, Stephen D. Krasner, ed., a special issue of *International Organization*, v. 36, n. 2 (Spring 1982).

BIBLIOGRAPHY

Acme-Cleveland. *Annual Reports*, various issues.

American Machinist, various issues.

American Metal Market/Metalworking News, various issues.

Angell, Sir Norman. *The Foundations of International Policy* (London: William Heinemann, 1914).

Aquino, Antonio. "Intra-Industry Trade and Inter-Industry Specialization as Concurrent Sources of International Trade in Manufactures," *Weltwirtschaftliches Archiv*, b. 114, h. 2, (1978).

Bachrach, Peter and Morton Baratz. "Decisions and Nondecisions: An Analytical Framework," *American Political Science Review*, 57 (1963).

Balance of Power or Hegemony: The Interwar Monetary System, Benjamin M. Rowland, ed. (New York: New York University Press, 1976).

Balassa, Bela. "The 'New Protectionism' and the International Economy," *Journal of World Trade Law*, v. 12, n. 5 (September-October 1978).

__________. "Tariff Protection in Industrial Countries: An Evaluation," *Journal of Political Economy*, v. 73, n. 6 (December 1965).

Balassa, Bela and Associates. *The Structure of Protection in Developing Countries* (Baltimore: Johns Hopkins Press, 1971).

Baldwin, David A. "Interdependence and Power: A Conceptual Analysis," *International Organization*, v. 34, n. 4 (Autumn 1980).

__________. "Power Analysis and World Politics: New Trends Versus Old Tendencies," *World Politics*, v. 31, n. 2 (January 1979).

Baldwin, Robert E. *The Multilateral Trade Negotiations: Toward Greater Liberalization?* (Washington: American Enterprise Institute, 1979).

__________. "The Political Economy of Postwar U.S. Trade Policy," *Bulletin, New York University Graduate School of Business*, No. 4, 1976.

__________. "Protectionist Pressures in the United States" in *Challenges to a Liberal International Economic Order* (Washington, D.C.: American Enterprise Institute, 1979).

Barker, Betty L. "U.S. Foreign Trade Associated with U.S. Multinational Companies," *Survey of Current Business*, v. 52, n. 12 (December 1972).

Barnet, Richard J. and Ronald E. Muller. *Global Reach: The Power of the Multinational Corporations* (New York: Simon and Schuster, 1974).

Batra, Ravundra and Francisco R. Cases. "Traded and Nontraded Intermediate Inputs, Real Wages, and Resource Allocation," *Canadian Journal of Economics*, v. 7, n. 2 (May 1974).

Bauer, Raymond A., Ithiel de Sola Pool, and Lewis Anthony Dexter. *American Business and Public Policy: The Politics of Foreign Trade* (Chicago: Aldine, Atherton, 1972).

Bendix. *Annual Reports*, various issues.

Bergsten, C. Fred, Robert O. Keohane, and Joseph S. Nye, Jr. "International Economics and International Politics: A Framework for Analysis," in *World Politics and International Economics*, Bergsten and Lawrence B. Krause, eds. (Washington, D.C.: Brookings Institution, 1975).

Bergsten, C. Fred, Thomas Horst and Theodore H. Moran. *American Multinationals and American Interests* (Washington: Brookings Institution, 1978).

Between Power and Plenty: Foreign Economic Policies of Advanced Industrial States, Peter J. Katzenstein, ed. (Madison: University of Wisconsin Press, 1978).

Blackhurst, Richard, Nicolas Marian and Jan Tumlir. *Trade Liberalization, Protectionism, and Interdependence*, GATT Studies in International Trade, n. 5 (Geneva: GATT, November 1977).

Blau, Peter M. *Exchange and Power in Social Life* (New York: Wiley, 1964).

Block, Fred L. *The Origins of International Economic Order: A Study of United States International Monetary Policy from World War II to the Present* (Berkeley: University of California Press, 1977).

Boothroyd, Geoffrey. *Fundamentals of Metal Machining and Machine Tools* (New York: McGraw-Hill, 1975).

Bryant, Ralph C. *Money and Monetary Policy in Interdependent Nations* (Washington: Brookings Institution, 1980).

Business Week, various issues.

Cable, Vincent. "Britain, The New Protectionism and Trade with the Newly Industrializing Countries," *International Affairs*, v. 55, n. 1(January 1979).

___________. "British Protectionism and LDC Imports," *Overseas Development Institute Review*, n. 2 (1977).

Calleo, David P. "Decline and Rebuilding of an International Economic System," in *Money and the Coming World Order*, David P. Calleo, ed. (York University Press, 1976).

Calleo David P. and Benjamin M. Rowland. *America and the World Political Economy: Atlantic Dreams and National Realities* (Bloomington: Indiana University Press, 1973).

Capitalism and the State in U.S.-Latin American Relations, Richard R. Fagen, ed. (Stanford: Stanford University Press, 1979).

James A. Caporaso, "Dependence, Dependency, and Power in the Global System: A Structural and Behavioral Analysis," *International Organization*, v. 32, n. 1 (Winter 1978).

Cardoso, Fernando Henrique and Enzo Faletto. *Dependency and Development in Latin America* (Berkeley: University of California Press, 1979).

Caves, Richard E. "Economic Models of Political Choice: Canada's Tariff Structure," *Canadian Journal of Economics*, v. 9 (1976).

Challenges to a Liberal International Economic Order (Washington, D.C.: American Enterprise Institute, 1979).

Change in the International System, Ole Holsti, Randolph M. Siverson, and Alexander L. George, eds. (Boulder: Westview Press, 1980).

Christensen, Cheryl. "Structural Power and National Security," in *Economic Issues and National Security*, Klaus Knorr and Frank N. Trager, eds. (Lawrence: The Regents Press of Kansas, 1977).

Cincinnati Milacron. *Annual Reports*, various issues.

Cooper, Richard N. "Economic Interdependence and Foreign Policy in the Seventies," *World Politics*, v. 25, n. 2 (January 1972).

___________. *The Economics of Interdependence: Economic Policy in the Atlantic Community* (New York: McGraw-Hill, 1968).

Coppcock, D.J. "Devaluation When Exports Have an Import Content," *Manchester School*, v. 39, n. 4 (December 1971).

Corden, W.M. "The Costs and Consequences of Protection: A Survey of Empirical Work" in *International Trade and Finance: Frontiers for Research*, Peter B. Kenen, ed. (Cambridge: Cambridge University Press, 1975).

__________. *The Theory of Protection* (London: Oxford University Press, 1974).

Cross, *Annual Reports*, various issues.

Cross & Trecker. *Annual Reports*, various issues.

Crozier, Michael, Samuel Huntington, and Joji Watanuki. *The Crisis of Democracy: Report on the Governability of Democracies to the Trilateral Commission* (New York: New York University Press, 1975).

Curzon, Gerard. "Neo-protectionism, the MFA and the European Community," *World Economy*, v. 4, n. 3 (September 1981).

Curzon, Gerard and Victoria Curzon. "The Management of Trade Relations in the GATT," in *International Economic Relations of the Western World 1959-1971*, Andrew Shonfield, ed. (Oxford: Oxford University Press, 1976).

D'Arista, Jane. "Private Overseas Lending: Too Far, Too Fast?" in *Debt and the Less Developed Countries*, Jonathan Aronson, ed. (Boulder: Westview, 1979).

Das, Satya P. "Traded Intermediate Products and the Theory of Devaluation," *International Economic Review*, v. 21, n. 2 (June 1980).

Debt and the Less Developed Countries, Jonathan Aronson, ed. (Boulder: Westview, 1979).

Destler, I.M. *Making Foreign Economic Policy* (Washington: Brookings Institution, 1980).

DosSantos, Theotonio. "The Structure of Dependence," *American Economic Review*, v. 60, n. 2 (May 1970).

Dun & Bradstreet, Dun's Marketing Services. *Million Dollar Directory 1982*.

Duvall, Raymond D. "Dependence and Dependencia Theory: Notes Toward Precision of Concept and Argument," *International Organization*, v. 32, n. 1 (Winter 1978).

Economic Issues and National Security, Klaus Knorr and Frank N. Trager, eds. (Lawrence: The Regents Press of Kansas, 1977).

Economic Stabilization in Developing Countries, William R. Cline and Sidney Weintraub, eds. (Washington: Brookings Institution, 1981).

Economist, various issues.

Effective Tariff Protection, H.G. Grubel and H.G. Johnson, eds. (Geneva: GATT and Graduate Institute of International Studies, 1971).

Elkan, Peter G. *The Meaning of Protection* (Wellington: N.Z. Institute of Economic Research, 1977).

Emerson, Richard M. "Power-Dependence Relations," *American Sociological Review*, v. 27, n. 1 (February 1962).

Evans, Peter. *Dependent Development: The Alliance of Multinational, State, and Local Capital in Brazil* (Princeton: Princeton University, 1979).

Ex-Cell-O. *Annual Reports*, various issues.

Falk, Richard A. *A Study of Future Worlds* (New York: Free Press, 1975).

Finger, J.M. "The Industry-Country Incidence of 'Less than Fair Value' Cases in U.S. Import Trade," *Quarterly Review of Economics and Business*, v. 21, n. 2 (Summer 1981).

F. Jos Lamb. *Annual Reports*, various issues.

Franko, Lawrence G. "Current Trends in Protectionism in Industrialized Countries: Focus on Western Europe," in *Protectionism or Industrial Adjustment?* G.K. Helleiner, et al., eds.(Paris: Atlantic Institute for International Affairs, April 1980).

Frieden, Jeff. "Third World Indebted Industrialization: International Finance and State Capitalism in Mexico, Brazil, Algeria, and South Korea," *International Organization*, v. 35, n. 3 (Summer 1981),p. 409.

The Future of the International Economic Order: An Agenda for Research, C. Fred Bergsten, ed. (Lexington, 1973).

Gard, Linda M. and James Riedel, "Safeguard Protection of Industry in Developing Countries," *Weltwirtschaftliches Archiv*, b. 116, h. 3(1980).

Gardner, Richard N. *Sterling-Dollar Diplomacy: Anglo-American Collaboration in the Reconstruction of Multilateral Trade* (New York: McGraw-Hill, 1969).

GATT, *The Tokyo Round of Multilateral Trade Negotiations: Volume II, Supplementary Report* (Geneva: GATT, 1980).

Giddings & Lewis. *Annual Reports*, various issues.

Gilpin, Robert. "Economic Interdependence and National Security" in *Economic Issues and National Security*, Klaus Knorr and Frank N. Trager, eds. (Lawrence: The Regents Press of Kansas, 1977).

__________. "The Politics of Transnational Economic Relations," in *Transnational Relations and World Politics*, Robert O. Keohane and Joseph S. Nye, eds., (Cambridge: Harvard University Press, 1972).

__________. *U.S. Power and the Multinational Corporations: The Political Economy of Foreign Direct Investment* (New York: Basic Books, 1975).

Goldstein, Judith L. "The State, Industrial Interests and Foreign Economic Policy: American Commercial Policy in the Postwar Period," (mimeo, n.d.).

Gray, H. Peter. *International Trade, Investment and Payments* (Boston: Houghton Mifflin, 1979).

Gray, James A. "The American Machine Tool Industry," February 12, 1982.

__________. "Machine Tool Exports: A Smaller Slice of a Bigger Pie," November 1979.

Grubel, Herbert G. and P.J. Lloyd, *Intra-Industry Trade: The Theory and Measurement of International Trade in Differentiated Products* (New York: John Wiley & Sons, 1975).

Haas, Ernst B. "Why Collaborate? Issue-Linkage and International Regimes," *World Politics*, v. 32, n. 3 (April 1980).

Haberler, Gottfried von. *The Theory of International Trade with its Application to Commercial Policy* (New York: Macmillan, 1936).

Hawkins, Robert G. "U.S. Multinational Investment in Manufacturing and Domestic Economic Performance," *Occasional Paper*, n. 1 (Washington: Center for Multinational Studies, 1972).

Helleiner, G.K. "The New Industrial Protectionism and the Developing Countries," in *Protectionism or Industrial Adjustment?* Helleiner, et al., eds.(Paris: Atlantic Institute for International Affairs, April 1980).

__________. "Transnational Enterprises and the New Political Economy of U.S. Trade Policy," *Oxford Economic Papers*, v. 29, n. 1 (March 1977).

Hesse, Helmut. "Hypotheses for the Expansion of Trade Between Industrial Countries, 1953-1970," in T*he International Division of Labor: Problems and Perspectives*, Herbert Giersch, ed. (Tubingen, 1974).

Hewer, Alan. "Manufacturing Industry in the Seventies: An Assessment of Import Penetration and Export Performance," *Economic Trends*, n. 320 (Great Britain, Central Statistical Office, June 1980).

Hindley, Brian. "VERs and GATT's Main Escape Clause," *World Economy*, v. 3, n. 3 (November 1980).

Hintz, Otto, et al. *Machine Tool Industry Final Report*, U.S. Army Industrial Base Engineering Activity, Rock Island, Illinois, November 1978.

Hirschman, Albert O. *National Power and the Structure of Foreign Trade* (Berkeley: University of California Press, 1945).

Hoffman, Stanley. "Notes on the Elusiveness of Modern Power," *International Journal*, 30 (Spring 1975).

Holsen, John and Jean Waelbroeck. *LDC Balance of Payments Policy and the International Monetary System*, World Bank Staff Working Paper, n. 226 (Washington: World Bank, 1976).

Holsti, Kal J. "A New International Politics? Diplomacy in Complex Interdependence," *International Organization*, v. 32, n. 2 (Spring 1978).

Houdaille Industries, Inc. *Petition to the President of the United States Through the Office of the United States Trade Representative for the Exercise of Presidential Discretion Authorized by Section 103.*

Hughes, Kent H. *Trade, Taxes, and Transnationals* (New York: Praeger, 1979).

Industry Week, various issues.

The International Corporation, Charles P. Kindleberger, ed. (Cambridge: MIT Press, 1970).

The International Division of Labor: Problems and Perspectives, Herbert Giersch, ed. (Tubingen, 1974).

International Economic Relations of the Western World 1959-1971, Andrew Shonfield, ed. (Oxford: Oxford University Press, 1976).

International Regimes, Stephen D. Krasner, ed., a special issue of *International Organization*, v. 36, n. 2 (Spring 1982).

International Trade and Finance: Frontiers for Research, Peter B. Kenen, ed. (Cambridge: Cambridge University Press, 1975).

Iron Age, various issues.

Japan Economic Journal, various issues.

Johnson, Harry G. *Aspects of the Theory of Tariffs* (Cambridge: Harvard University Press, 1971).

Katzenstein, Peter J. "Capitalism in One Country?: Switzerland in the International Economy," *International Organization*, v. 34, n. 4 (Autumn 1980).

__________. "Conclusion: Domestic Structures and Strategies of Foreign Economic Policy" in *Between Power and Plenty: Foreign Economic Policies of Advanced Industrial States*, Katzenstein, ed. (Madison: University of Wisconsin Press, 1978).

__________. "International Interdependence: Some Long-Term Trends and Recent Changes," *International Organization*, v. 29, n. 4(Autumn 1975).

__________. "International Relations and Domestic Structures: Foreign Economic Policies of Advanced Industrial States," *International Organization*, v. 30, n. 1 (Winter 1976).

Kearney & Trecker. *Annual Reports*, various issues.

Keohane, Robert O. "The Theory of Hegemonic Stability and Changes in International Economic Regimes, 1967-1977," in *Change in the International System*, Ole Holsti, Randolph M. Siverson, and Alexander L. George, eds. (Boulder: Westview Press, 1980).

Keohane, Robert O. and Joseph S. Nye, *Power and Interdependence: World Politics in Transition* (Boston: Little, Brown, 1977).

__________. "Transnational Relations and World Politics: An Introduction," and "A Conclusion," in *Transnational Relations and World Politics*, Keohane and Nye, eds., (Cambridge: Harvard University Press, 1972).

__________. "World Politics and the International Economic Systems" in *The Future of the International Economic Order: An Agenda for Research*, C. Fred Bergsten, ed. (Lexington, 1973).

Kindleberger, Charles P. "Systems of International Economic Organization," in *Money and the Coming World Order*, David P. Calleo, ed. (York University Press, 1976).

__________. *The World in Depression, 1929-1939* (Berkeley: University of California Press, 1973).

Knorr, Klaus. "Economic Interdependence and National Security," in *Economic Issues and National Security*, Knorr and Frank N. Trager, eds. (Lawrence: The Regents Press of Kansas, 1977).

__________. "International Leverage and Its Uses" in *Economic Issues and National Security*, Knorr and Frank N. Trager, eds. (Lawrence: The Regents Press of Kansas, 1977).

__________. "National Power in an Economically Interdependent World," in *Power, Capabilities, Interdependence: Problems in the Study of International Influence*, Kjell Goldman and Gunnar Sjostedt, eds.(London: Sage Publications, 1979).

__________. *The Power of Nations: The Political Economy of International Relations* (New York: Basic Books, 1975).

Krasner, Stephen D. *Defending the National Interest: Raw Materials Investments and U.S. Foreign Policy* (Princeton: Princeton University Press, 1979).

__________. "State Power and the Structure of International Trade," *World Politics*, v. 28, n. 3 (April 1978).

__________. "The Tokyo Round: Particularistic Interests and Prospects for Stability in the Global Trading System," *International Studies Quarterly*, v. 23, n. 4 (December 1979).

__________. "U.S. Commercial and Monetary Policy: Unraveling the Paradox of External Strength and Internal Weakness," in *Between Power and Plenty: Foreign Economic Policies of Advanced Industrial States*, Peter J. Katzenstein, ed. (Madison: University of Wisconsin Press, 1978).

Krause, Lawrence. "Private International Finance," in *Transnational Relations and World Politics*, Robert O. Keohane and Joseph S.. Nye, eds., (Cambridge: Harvard University Press, 1972).

Krauss, Melvyn. *The New Protectionism: The Welfare State and International Trade* (New York: New York University Press, 1978).

Leith, J. Clark. "The Effects of Tariffs on Production, Consumption and Trade: A Revised Analysis," *American Economic Review*, v. 61, n. 1(March 1971).

Lipson, Charles. "The IMF, Commercial Banks and Third World Debts," in *Debt and the Less Developed Countries*, Jonathan Aronson, ed. (Boulder: Westview, 1979).

Litton. *Annual Reports*, various issues.

Maddigan, Ruth J. "The Measurement of Vertical Integration," *Review of Economics and Statistics*, v. 63, n. 3 (August 1981).

McKeown, Timothy J. "Hegemonic Stability and Nineteenth Century Tariff Levels in Europe," *International Organization*, v. 37, n. 1 (Winter 1983).

McKinnon, Ronald I. "Intermediate Products and Differential Tariffs: A Generalization of Lerner's Symmetry Theorem," *Quarterly Journal of Economics*, v. 80, n. 4 (November 1966).

Michaely, Michael. *Theory of Commercial Policy: Trade and Protection* (Chicago: University of Chicago Press, 1977).

__________. "The Welfare Costs of Negative Value-Added," *Journal of International Economics*, v. 5, n. 3 (August 1975).

Money and the Coming World Order, David P. Calleo, ed. (York University Press, 1976).

Moran, Theodore H. *Multinational Corporations and the Dependence: Copper in Chile* (Princeton: Princeton University Press, 1974).

Monroe, Wilbur F. *Japanese Exports to the United States: Analysis of Import Pull and Export Push Factors* (Washington: U.S.-Japan Trade Council, June 2, 1978).

Morkre, Morris E. and David G. Tarr. *The Effects of Restrictions on United States Imports: Five Case Studies and Theory*, Staff Report of the Bureau of Economics to the Federal Trade Commission (Washington: Government Printing Office, June 1980).

Morse, Edward L. *Modernization and the Transformation of International Relations* (New York: Free Press, 1976).

__________. "The New Economic Nationalism and the Coordination of Economic Policies," in *The New Nationalism: Implications for Transatlantic Relations*, Werner Link and Werner J. Feld, eds. (New York: Pergamon Press, 1979).

__________. "The Transformation of Foreign Policies: Modernization, Interdependence, and Externalization," *World Politics*, v. 22, n. 3(April, 1970).

__________. "Transnational Economic Processes," in *Transnational Relations and World Politics*, Robert O. Keohane and Joseph S.. Nye, eds., (Cambridge: Harvard University Press, 1972).

Murdock, Clark A. "Economic Factors as Objects of Security," in *Economic Issues and National Security*, Klaus Knorr and Frank N. Trager, eds. (Lawrence: The Regents Press of Kansas, 1977).

Murray, Tracy, et al., "Alternative Forms of Protection Against Market Disruption," *Kyklos*, v. 31, F. 4 (1978).

National Machine Tool Builders' Association. *Directory 1982.*

__________. "Industry Estimates of Machine Tool New Orders, Cancellations, Shipments and Backlog," n.d.

__________. *1981-1982 Economic Handbook the Machine Tool Industry*.

National Science Foundation. "National Patterns of Science and Technology Resources, 1980," NSF 80-308.

The New Nationalism: Implications for Transatlantic Relations, Werner Link and Werner J. Feld, eds. (New York: Pergamon Press, 1979).

New York Times, various issues.

Nowzad, Bahram. *The Rise in Protectionism* (Washington: International Monetary Fund, 1978).

O'Donnell, Guillermo. *Modernization and Bureaucratic-Authorization: Studies in South American Politics* (Berkeley: Institute of International Studies, 1973).

__________. "Reflections on the Patterns of Change in the Bureaucratic-Authoritarian State," *Latin American Research Review*, v. 13, n. 1 (1978).

OECD. *National Accounts of OECD Countries, Volume II: Detailed Tables: Statistics of Foreign Trade, Annual: Tables by Reporting Countries* (Series B).

__________. *Statistics of Foreign Trade, Annual: Tables by Reporting Countries*.

Pagoulatos, Emilio and Robert Sorensen. "Two-Way International Trade: An Econometric Analysis," *Weltwirtschaftliches Archiv*, b. 111, h. 3

Pastor, Robert A. *Congress and the Politics of U.S. Foreign Economic Policy: 1929-1976* (Berkeley: University of California Press, 1980).

Payer, Cheryl. *The Debt Trap: The IMF and the Third World* (New York: Monthly Review, 1974).

Pincus, J.J. "Pressure Groups and the Pattern of Tariffs," *Journal of Political Economy*, v. 83, n. 4 (August 1978).

Power, Capabilities, Interdependence: Problems in the Study of International Influence, Kjell Goldman and Gunnar Sjostedt, eds.(London: Sage Publications, 1979).

Protectionism or Industrial Adjustment? G.K. Helleiner, et al., eds.(Paris: Atlantic Institute for International Affairs, April 1980).

Ray, Edward John. "The Determinants of Tariff and Non-Tariff Trade Restrictions in the United States," *Journal of Political Economy*, v. 89, n. 1 (February 1981).

___________. "The Optimum Commodity Tariff and Tariff Rates in Developed and Less Developed Countries," *Review of Economics and Statistics*, v. 56 (August 1974).

Real, Bernard. *Technical Change and Economic Policy: Science and Technology in the New Economic and Social Contest, Sector Report, The Machine Tool Industry* (Paris: OECD, August 1, 1980).

Reppy, Judith. "The Role of the Air Force in the Development of Numerically-Controlled Machine Tools," 1981 (mimeo).

"Report of the Industry Sector Advisory Committee on Machine Tools, Other Metalworking Equipment and Other Nonelectrical Machinery (ISAC 17) to the President, the Congress, and the Special Representative for Trade Negotiations, April 18, 1979," in U.S., Congress, Senate, Committee on Finance, Subcommittee on International Trade, *Private Advisory Committee Reports on the Tokyo Round of Multilateral Trade Negotiations*, August 1979.

Richardson, Neil R. *Foreign Policy and Economic Dependence* (Austin: University of Texas Press, 1973).

Rosecrance, Richard. "The One World of Hans Morgenthau," *Social Research*, v. 48, n. 4 (Winter 1981).

Rosecrance, Richard and Arthur Stein. "Interdependence: Myth or Reality?" *World Politics*, v. 26, n. 1 (October 1973).

Rosecrance, Richard, et al., "Whither Interdependence?" *International Organization*, v. 31, n. 3 (Summer 1977).

Ruffin, Roy J. "Tariffs, Intermediate Goods, and Domestic Production," *American Economic Review*, v. 59, n. 3 (June 1969).

Sampson, Gary P. "Contemporary Protectionism and the Exports of Developing Countries,: *World Development*, V 8 (Fall 1979).

Schattschneider, E.E. *Politics, Pressures and the Tariff: A Study of Free Enterprise Politics as Shown in the 1929-30 Revision of the Tariff* (Englewood Cliffs, N.J.: Prentice-Hall, 1935).

Shea, Koon-Lum. "Imported Inputs, Devaluation and Balance of Payments: A Keynesian Macro-Approach," *Southern Economic Journal*, v. 43,n. 2 (October 1976).

Solomon, Robert. *The International Monetary System, 1945-1976: An Insider's View* (New York: Harper & Row, 1977).

Stallings, Barbara. "Peru and the U.S. Banks: Privatization of Financial Relations," in *Capitalism and the State in U.S.-Latin American Relations*, Richard R. Fagen, ed. (Stanford: Stanford University Press, 1979).

Standard & Poor's Industry Survey, various issues.

"Statement by Jack E. Moran, Chairman of the Board and Chief Executive Officer, the Carlton Machine Tool Company, before the Joint Economic Committee, September 30, 1981," (mimeo).

"Statement by National Machine Tool Builders' Association before the Subcommittee on Trade, Committee on Ways and Means, U.S. House of Representatives, December 15, 1981," (mimeo).

"Statement by Wayne R. Moore, President and Chief Executive Officer, Moore Special Tool Company, Inc. before the Subcommittee on Banking, Housing, and Urban Affairs, U.S. House of Representatives, April 22, 1982,"(mimeo).

Steel, various issues.

Stepan, Alfred. *The State and Society: Peru in Comparative Perspective* (Princeton: Princeton University Press, 1978).

Strange, Susan. "International Monetary Relations," in *International Economic Relations of the Western World 1959-1971*, Andrew Shonfield, ed. (Oxford: Oxford University Press, 1976), Volume 2.

__________. "The Management of Surplus Capacity: Or How Does Theory Stand Up to Protectionism 1970s Style?" *International Organization*, v. 33, n. 3 (Summer 1979).

Szentes, Tamas. *The Political Economy of Underdevelopment* (Budapest: Akademia Kiado, 1971).

Textron. *Annual Reports*, various issues.

Transnational Relations and World Politics, Robert O. Keohane and Joseph S. Nye, eds., (Cambridge: Harvard University Press, 1972).

Travis, William Penfield. *The Theory of Trade and Protection* (Cambridge: Harvard University Press, 1964).

Tsoukalis, Loukas and Antonio da Silva Ferreira. "Management of Industrial Surplus Capacity in the European Community," *International Organization*, v. 34, n. 3 (Summer 1980).

Tucker, Irvin B. and Ronald P. Wilder. "Trends in Vertical Integration in the U.S. Manufacturing Sector," *Journal of Industrial Economics*, v. 26, n. 1 (September 1977).

U.N , Statistical Office. *Monthly Bulletin of Statistics*, various issues.

__________. *Yearbook of Industrial Statistics*, various issues.

__________. *Yearbook of International Trade Statistics*, various issues.

UNCTAD. *Handbook of International Trade Statistics*, various issues.

U.S., Bureau of Economic Analysis. *Business Statistics*, various issues.

__________. *Survey of Current Business*, various issues.

U.S., Bureau of Labor Statistics. *Occupational Employment Statistics*, various issues.

__________. *Producer Prices and Price Indexes*, various issues.

__________. *Survey of Manufacturing*, various issues.

U.S., Bureau of the Census. *Annual Survey of Manufactures: Statistics for Industry Groups and Industries*, various issues.

___________. *Annual Survey of Manufactures: Value of Product Shipments*, various issues.

___________. *Census of Manufactures*, various issues.

___________. *Commodity Exports and Imports Related to Output*, various issues.

___________. "Current Industrial Reports, Series MQ-35W, Metalworking Machinery,"various issues.

___________. *Numerical List of Manufactured Products*, October 1978.

___________. *1977 Census of Manufactures: Industry Series, Metalworking Machinery and Equipment*, June 1980.

___________. *1977 Enterprise Statistics: General Report on Industrial Organization*, April 1981.

___________. *U.S. Commodity Exports and Imports as Related to Output*, ES2, various issues.

___________. *U.S. Exports: Domestic Merchandise, SIC-Based Products by World Area*, FT 610, various issues.

___________. *U.S. Exports: Schedule B Section by Division by Group by Number by Country of Destination by Custom District of Exportation and Method of Transportation*, EM 522, various issues.

___________. *U.S. Exports: SIC-Based Products by World Area, FT 610 and U.S. Imports: SIC-Based Products by World Area*, FT 210, various issues.

___________. *U.S. Exports: SIC Division by SIC-Based Two-Digit, Three-Digit and Four-Digit Product Code*, various issues.

___________. *U.S. General Imports of Cotton Manufactures: Country and Geneva Agreement Category*, FT 130, various issues.

___________. *U.S. Imports: Consumption and General SIC-Based Products by World Areas*, FT 210, various issues.

___________. *U.S. Imports for Consumption: TSUSA Schedule by TSUSA Number by Unit Control by Country of Origin*, IM 146, various issues.

___________. *U.S. Imports: SIC Division by SIC-Based Two-Digit, Three-Digit, and Four-Digit Product Code*, IA 275, various issues.

U.S., Comptroller General. "Followup on Use of Numerically Controlled Equipment to Improve Defense Plant Productivity," LCD-78-427, January 11, 1979.

U.S., Congress, House of Representatives, Committee on Banking and Currency. *Export-Import Bank Act Extension*, Hearings, February 25, 1963, 88th Congress, First Session.

U.S., Congress, House of Representatives, Committee on Banking and Currency, Subcommittee on International Trade. *International Economic Policy*, Hearings, April 22, 1974, 93rd Congress, Second Session.

U.S., Congress, House of Representatives, Committee on Banking, Finance and Urban Affairs and Committee on International Trade, Investment and Monetary Policy. *Export-Import Bank and Trade of China*, Hearings, January 26, 1978, 95th Congress, Second Session.

__________. *Export Trading Companies Legislation.* Hearings. July 1, 1980, 96th Congress, Second Session.

U.S., Congress, House of Representatives, Committee on Education and Labor, Subcommittee on the Impact of Imports and Exports on American Employment. *Impact of Imports and Exports on Employment*, Hearings, Part 8, November 27, 1961, 87th Congress, First and Second Sessions.

U.S., Congress, House of Representatives, Committee on Foreign Affairs, Subcommittee on International Economic Policy and Trade. *Extension and Revision of the Export Administration Act of 1969*, Hearings and Markup, February 15, 1979, 96th Congress, First Session.

U.S., Congress, House of Representatives, Committee on International Relations. *Extension of the Export Administration Act of 1969*, Hearings, Part 1, June 8, 1976, 94th Congress, Second Session.

U.S., Congress, House of Representatives, Committee on International Relations, Subcommittee on International Economic Policy and Trade. *Export Licensing: Cocom List Review Proposals of the United States*, Hearings, June 14, 1978, 95th Congress, Second Session.

U.S., Congress, House of Representatives, Committee on Ways and Means. *Advisability of a Tax Reduction in 1980 Effective for 1981*, Hearings, Part 1, July 22,1980, 96th Congress, Second Session.

__________. *Foreign Trade and Tariff Proposals*, Hearings, Part 7, June 21, 1968, 90th Congress, Second Session.

__________. *Renewal of the Trade Agreements Act*, Hearings, Part 2, March 10, 1958, 85th Congress, Second Session.

__________. *Tariff and Trade Proposals*, Hearings, Part 9, June 4, 1970, 91st Congress, Second Session.

__________. *Tariff Readjustment - 1929*, Hearings, v. 3, Schedule 3, January 14, 1929, 70th Congress, Second Session.

__________. *Tax Aspects of the President's Economic Program*, Hearings, Part 2, February 24, 1981, 97th Congress, First Session.

__________. *Trade Agreements Extension*, Hearings, Part 2, January 1?, 1955, 84th Congress, First Session.

__________. *Trade Expansion Act of 1962*, Hearings, Part 2, March 12, 1962, 87th Congress, Second Session.

__________. *Trade Reform*, Hearings, Part 3, May 14, 1973, 93rd Congress, First Session.

U.S., Congress, House of Representatives, Committee on Ways and Means, Subcommittee on Trade. *Background Material on the Multifiber Arrangement*, Hearing, 96th Congress, 1st Session, April 2, 1979.

__________. *Causes and Consequences of the U.S. Trade Deficit and Developing Problems in U.S. Exports*, Hearings, November 11, 1977, 95th Congress, First Session.

__________. *Trade with Japan*, Hearings, August 26, 1980, 96th Congress, Second Session.

__________. *United States-China Trade Agreement*, Hearings, November 1, 1979, 96th Congress, First Session.

U.S., Congress, House of Representatives, Select Committee on Export Control. *Investigation and Study of the Administration, Operation, and Enforcement of the Export Control Act of 1949, and Related Acts*, Hearings, November 25,1961, 87th Congress, First Session.

U.S., Congress, Senate. *Implications of Multinational Firms for World Trade and Investment and for U.S. Trade and Labor*, by U.S. Tariff Commission, February 1973.

U.S., Congress, Senate. Committee on Banking and Currency. *Export-Import Bank Act Amendments of 1954*, Hearings, June 14, 1954, 83rd Congress, Second Session.

U.S., Congress, Senate, Committee on Banking, Housing and Urban Affairs, Subcommittee on International Finance. *Amend the Export-Import Bank Act of 1945*, Hearings, March 8, 1971, 92nd Congress, First Session.

__________. *Amending the Export-Import Bank Act and the Trade Act*, Hearings, July 12, 1979, 96th Congress, First Session.

__________. *Authority for Regulation of Exports - 1972*, Hearings, May 13, 1972, 92nd Congress, Second Session.

__________. *Export-Import Bank of the United States*, Hearings, October 29, 1973, 93rd Congress, First Session.

__________. *Export Policy*, Hearings, Part 4, March 20, 1978, 95th Congress, Second Session.

__________. *Extension of the Export Administration Act*, Hearings, March 2, 1976, 94th Congress, Second Session.

__________. *U.S. Export Control Policy and Extension of the Export Administration Act*, Hearings, Part 2, March 12, 1979, 96th Congress, First Session.

U.S., Congress, Senate, Committee on Banking, Housing and Urban Affairs, Subcommittee on International Finance and Monetary Policy. *Export Trading Company Act of 1981*, Hearings, February 17, 1981, 97th Congress, First Session.

U.S., Congress, Senate, Committee on Banking, Housing and Urban Affairs, Subcommittee on International Trade. *Export Trading Company Act of 1980*, March 17, 1980, 96th Congress, Second Session.

U.S., Congress, Senate, Committee on Finance. *Trade Expansion Act of 1962*, Hearings, Part 3, August 7, 1962, 87th Congress, Second Session.

__________. *The Trade Reform Act of 1973*, Hearings, Part 3, March 26, 1974, 93rd Congress, Second Session.

U.S., Congress, Senate, Committee on Finance, Subcommittee on International Trade. *Agreement on Trade Relations Between the United States and the People's Republic of China*, Hearings, November 15, 1979, 96th Congress, First Session.

__________. *Private Advisory Committee Reports on the Tokyo Round of Multilateral Trade Negotiations*, August 1979.

__________. *U.S. International Trade Strategy*, Hearing, July 28, 1980, 96th Congress, 2d Session.

U.S., Congress, Senate, Committee on Finance, Subcommittee on Taxation and Debt Management Generally. *Tax Incentives for Exports*, Hearings, June 18, 1979, 96th Congress, First Session.

U.S., Congress, Senate, Select Committee on Small Business. *Machine Tool Programs*, Hearings, February 7, 1956, 84th Congress, Second Session.

__________. *Machine Tool Shortages*, Hearings, February 27, 1952, 82nd Congress, Second Session.

__________. *Small Business Participation in U.S. Exports*, Hearings, September 25, 1979, 96th Congress, First Session.

U.S., Department of Commerce, International Trade Administration. *International Economic Indicators*, various issues.

U.S., Department of State. *Treaties and Other International Acts Series*, various issues.

__________. *Treaties in Force*, various issues.

__________. *United States Treaties and Other International Agreements*, various issues.

U.S., Department of the Treasury. *Customs Bulletin*, various issues.

__________. *Customs Bulletin and Decisions*, various issues.

U.S. *Federal Register*, various issues.

U.S., Federal Trade Commission. "Quarterly Financial Report for Manufacturing Corporations, " various issues.

U.S., Internal Revenue Service. *Corporation Sourcebook of Statistics of Income*, various issues.

U.S., International Trade Administration. *International Economic Indicators*, various issues.

U.S., International Trade Commission. *Annual Reports*, various issues.

__________. *Certain Motor Vehicles and Chassis and Bodies Therefor: Report to the President on Investigation No. TA-201-44 Under Section 201 of the Trade Act of 1974* (USITC Publication 1110, December 1980).

__________. *Clothespins: Report to the President on Investigation No. TA-201-36 Under Section 291 of the Trade Act of 1974* (USITC Publication 933, December 1978).

__________. *Color Television Receivers and Subassemblies Thereof: Report to the President on Investigation No. TA-203-b Under Section 203 of the Trade Act of 1974* (USITC Publication 1068, May 1980).

__________. *Footwear: Report to the President on Investigation No. TA-201-7 Under Section 201 of the Trade Act of 1974* (USITC Publication 758, February 1976).

__________. *Footwear: Report to the President on Investigation No. TA-201-18 Under Section 201 of the Trade Act of 1974*.

__________. *The History and Current Status of The Multifiber Arrangement* (USITC Publication 850, January 1978).

__________. *The History and Current Status of the Multifiber Arrangement: The Multifiber Arrangement, 1973 to 1980*. Report on Investigation No. 332-108 Under Section 332 of the Tariff Act of 1930, v. 1 (USITC Publication 1131, March 1981).

__________. *Live Cattle and Certain Edible Meat Products of Cattle: Report to the President on Investigation No. TA-201-25 Under Section 201 of the Trade Act of 1974* (USITC Publication 834, September 1977).

__________. *Mushrooms: Report to the President on Investigation No. TA-201-17 Under Section 291 of the Trade Act of 1974* (USITC Publication 798, January 1977).

__________. *Mushrooms: Report to the President on Investigation No. TA-201-43 Under Section 201 of the Trade Act of 1974* (USITC Publication 1089, August 1980).

__________. *Operation of the Trade Agreements Program*, various issues.

__________. *Stainless Steel and Alloy Tool Steel: Report to the President on Investigation No. TA-201-5 Under Section 201 of the Trade Act of 1974* (USITC Publication 756, January 1976).

__________. *Stainless Steel and Alloy Tool Steel: Report to the President on Investigation No. TA-203-3 Under Section 203(i)(2) of the Trade Act of 1974* (USITC Publication 838, October 1977).

__________. *Stainless Steel and Alloy Tool Steel: Report to the President on Investigation No. TA-203-5 Under Section 203 of the Trade Act of 1974* (USITC Publication 968, April 1979).

__________. *Summary of Trade and Tariff Information: Metalworking Machine Tools*, May 1981.

__________. *Television Receivers, Color and Monochrome, Assembled or Not Assembled, Finished or Not Finished, No. TA-201-19 Under Section 201(b) of the Trade Act of 1974* (USITC Publication 808, March 1977).

U.S., Office of the United States Trade Representative. *Trade Action Monitoring System.*

U.S., President. *Annual Report of the President of the United States on the Trade Agreements Program*, various issues.

U.S., Tariff Commission. *Operation of the Trade Agreement Program, 1963-64.*

The Value Line Investment Survey, various issues.

Walter, Ingo and Jae W. Chung. "The Pattern of Non-Tariff Obstacles to International Market Access," *Weltwirtschaftliches Archiv*, b. 108, h. 1 (1972).

Waltz, Kenneth N. *Man, the State and War: A Theoretical Analysis* (New York: Columbia University Press, 1959).

__________. "The Myth of National Interdependence," in *The International Corporation*, Charles P. Kindleberger, ed. (Cambridge: MIT Press, 1970).

__________. *Theory of International Politics* (Reading: Addison-Wesley, 1979).

Warner & Swasey. *Annual Reports*, various issues.

Washington Post, various issues.

Wells, J.D. and J.C. Imber. "The Home and Export Performance of United Kingdom Industries," *Economic Trends*, n. 287 (Great Britain, Central Statistical Office, August 1917).

Williams, James R. *The Canadian-United States Tariff and Canadian Industry: A Multisectoral Analysis* (Toronto: University of Toronto Press, 1979).

World Politics and International Economics, C. Fred Bergsten and Lawrence B. Krause, eds. (Washington, D.C.: Brookings Institution, 1975).

Yeager, Leland B. *International Monetary Relations: Theory, History and Policy* (New York: Harper & Row, 1976).

Young, Oran R. "International Regimes: Problems of Concept Formation," *World Politics*, v. 32, n. 3 (April 1980).

Index